BIOCHEMISTRY RESEARCH TREND

THE THIRD ELEMENT - ACTINIDIC ARCHAEA, DIGOXIN AND THE BIOLOGICAL UNIVERSE

BIOCHEMISTRY RESEARCH TREND

Additional books in this series can be found on Nova's website under the Series tab.

Additional e-books in this series can be found on Nova's website under the E-book tab.

BIOCHEMISTRY RESEARCH TREND

THE THIRD ELEMENT - ACTINIDIC ARCHAEA, DIGOXIN AND THE BIOLOGICAL UNIVERSE

RAVIKUMAR KURUP
AND
PARAMESWARA ACHUTHA KURUP

Nova Science Publishers, Inc.
New York

For permission to use material from this book please contact us:
Telephone 631-231-7269; Fax 631-231-8175
Web Site: http://www.novapublishers.com

NOTICE TO THE READER

Library of Congress Cataloging-in-Publication Data

The third element : actinidic archaea, digoxin, and the biological universe / [edited by] Ravikumar A. Kurup and Parameswara Achutha Kurup.
p. cm.
Includes index.
ISBN 978-1-61942-318-3 (hardcover)
1. Archaebacteria--Molecular aspects. 2. Viroids. 3. Digoxin. 4. Molecular evolution. 5. Actinide elements--Environmental aspects. I. Kurup, Ravikumar. II. Kurup, Parameswaran Achutha.
QR82.A69T45 2011
579.3'21--dc23 2011052620

Published by Nova Science Publishers, Inc. † New York

Contents

Preface

An actinide dependent shadow biosphere of archaea and viroids in disease states is described with archaeal digoxin playing a pathological role. Archaeal digoxin can produce neuro-immuno-endocrine-genetic-metabolic integration. Archaeal digoxin by producing sodium-potassium ATPase inhibition can generate synchronized sustained burst activity in neuronal cells resulting in a electromagnetic field and a plasma state in the cell. This can generate cellular and molecular electromagnetic traces important in signal transduction and quantal perception. Metal actinides in beach sands have been postulated to play a role in abiogenesis. A hypothesis of cholesterol as the primal prebiotic molecule synthesised on actinide surfaces with all other biomolecules arising from it and a self replicating cholesterol lipid organism as the initial life form is presented. A cholesterol based theory of abiogenesis and its role in the evolution of universe is elucidated.

Chapter I

A Cholesterol and Actinide Dependent Shadow Biosphere of Archaea and Viroids

Introduction

Endomyocardial fibrosis along with the root wilt disease of coconut is endemic to Kerala with its radioactive actinide beach sands. Actinides like rutile, endogenous digoxin as well as organisms like phytoplasmas and viroids have been implicated in the etiology of these diseases. [1,2,3,4]

Endogenous digoxin has been related to the pathogenesis of schizophrenia, malignancy, metabolic syndrome x, autoimmune disease and neuronal degeneration [4]. The possibility of endogenous digoxin synthesis by actinide based primitive organism like archaea with a mevalonate pathway and cholesterol catabolism was considered.[5-8]

An actinide dependent shadow biosphere of archaea and viroids in the above mentioned disease states is described.[7,9] Metal actinides in beach sands have been postulated to play a role in abiogenesis [7] A hypothesis of cholesterol as the primal prebiotic molecule synthesised on actinide surfaces with all other biomolecules arising from it and a self replicating cholesterol lipid organism as the initial life form is presented.

Materials and Methods

The following groups were included in the study:- endomyocardial fibrosis, alzheimer's disease, multiple sclerosis, non-hodgkin's lymphoma, metabolic syndrome x with cerebrovascular thrombosis and coronary artery disease, schizophrenia, autism, seizure disorder, creutzfeldt jakob disease and acquired immunodeficiency syndrome. There were 10 patients in each group and each patient had an age and sex matched healthy control selected randomly from the general population. The blood samples were drawn in the fasting state before treatment was initiated. Plasma from fasting heparinised blood was used and the experimental protocol was as follows (I) Plasma+phosphate buffered saline, (II) same as

I+cholesterol substrate, (III) same as II+rutile 0.1 mg/ml, (IV) same as II+ciprofloxacine and doxycycline each in a concentration of 1 mg/ml. Cholesterol substrate was prepared as described by Richmond [10] Aliquots were withdrawn at zero time immediately after mixing and after incubation at 37°C for 1 hour. The following estimations were carried out:- Cytochrome F420, free RNA, free DNA, polycyclic aromatic hydrocarbon, hydrogen peroxide, dopamine, serotonin, pyruvate, ammonia, glutamate, cytochrome C, hexokinase, ATP synthase, HMG CoA redutase, digoxin and bile acids.[11-13] Cytochrome F420 was estimated flourimetrically (excitation wavelength 420 nm and emission wavelength 520 nm). Polycyclic aromatic hydrocarbon was estimated by measuring hydrogen peroxide liberated by using glucose reagent. Informed consent of the subjects and the approval of the ethics committee were obtained for the study. The statistical analysis was done by ANOVA.

Results

Plasma of control subjects showed increased levels of the above mentioned parameters with after incubation for 1 hour and addition of cholesterol substrate resulted in still further significant increase in these parameters. The plasma of patients showed similar results but the extent of increase was more. The addition of antibiotics to the control plasma casued a decrease in all the parameters while addition of rutile increased their levels. The addition of antibiotics to the patient's plasma caused a decrease in all the parameters while addition of rutile increased their levels but the extent of change was more in patient's sera as compared to controls. The results are expressed in tables 1-7 as percentage change in the parameters after 1 hour incubation as compared to the values at zero time.

Discussion

There was increase in cytochrome F420 indicating archaeal growth. The archaea can synthesise and use cholesterol as a carbon and energy source. [6,14] The archeal origin of the enzyme activities was indicated by antibiotic induced suppression. The study indicates the presence of actinide based archaea with an alternate actinide based enzymes or metalloenzymes in the system as indicated by rutile induced increase in enzyme activities. [15] There was also an increase in archaeal HMG CoA reductase activity indicating increased cholesterol synthesis by the archaeal mevalonate pathway. The archaeal beta hydroxyl steroid dehydrogenase activity indicating digoxin synthesis and archaeal cholesterol hydroxylase activity indicating bile acid synthesis were increased.[8] The archaeal cholesterol oxidase activity was increased resulting in generation of pyruvate and hydrogen peroxide.[14] The pyruvate gets converted to glutamate and ammonia by the GABA shunt pathway. The archaeal aromatization of cholesterol generating PAH, serotonin and dopamine was also detected [16] The archaeal glycolytic hexokinase activity and archaeal extracellular ATP synthase activity were increased. The archaea can undergo magnetite and calcium carbonate mineralization and can exist as calcified nanoforms.[17] There was an increase in free RNA indicating self replicating RNA viroids and free DNA indicating generation of viroid complementary DNA strands by archaeal reverse transcriptase activity.

Table 1. Effect of rutile and antibiotics on cytochrome F420 and PAH

Group	CYT F420 % (Increase with Rutile)		CYT F420 % (Decrease with Doxy+Cipro)		PAH % change (Increase with Rutile)		PAH % change (Decrease with Doxy+Cipro)	
	Mean	± SD	Mean	± SD	Mean	± SD	Mean	± SD
Normal	4.48	0.15	1[illegible].24	0.66	4.45	0.14	18.25	0.72
Schizo	23.24	2.01	5[illegible].72	7.08	23.01	1.69	59.49	4.30
Seizure	23.46	1.87	5[illegible].27	8.86	22.67	2.29	57.69	5.29
AD	23.12	2.00	5[illegible].90	6.94	23.26	1.53	60.91	7.59
MS	22.12	1.81	6[illegible].33	9.82	22.83	1.78	59.84	7.62
NHL	22.79	2.13	5[illegible].90	7.29	22.84	1.42	66.07	3.78
DM	22.59	1.86	5[illegible].05	8.45	23.40	1.55	65.77	5.27
AIDS	22.29	1.66	5[illegible].02	7.50	23.23	1.97	65.89	5.05
CJD	22.06	1.61	5[illegible].81	6.04	23.46	1.91	61.56	4.61
Autism	21.68	1.90	5[illegible].93	9.64	22.61	1.42	64.48	6.90
EMF	22.70	1.87	6[illegible].46	8.06	23.73	1.38	65.20	6.20
	F value 306.749 P value < 0.001		F value 130.054 P value < 0.001		F value 391.318 P value < 0.001		F value 257.996 P value < 0.001	

Table 2. Effect of rutile and antibiotics on free RNA and DNA

Group	DNA % change (Increase with Rutile)		DNA % change (Decrease with Doxy+Cipro)		RNA % change (Increase with Rutile)		RNA % change (Decrease with Doxy+Cipro)	
	Mean	± SD	Mean	± SD	Mean	± SD	Mean	± SD
Normal	4.37	0.15	18.39	0.38	4.37	0.13	18.38	0.48
Schizo	23.28	1.70	51.41	3.36	23.59	1.83	65.69	3.94
Seizure	23.40	1.51	53.68	4.66	23.08	1.87	65.09	3.48
AD	23.52	1.65	54.15	4.60	23.29	1.92	65.39	3.95
MS	22.62	1.38	53.82	5.53	23.29	1.98	67.46	3.96
NHL	22.42	1.99	51.14	3.47	23.78	1.20	66.90	4.10
DM	23.01	1.67	55.35	3.56	23.33	1.86	66.46	3.65

Table 2. (Continued)

Group	DNA % change (Increase with Rutile)		DNA % change (Decrease with Doxy+Cipro)		RNA % change (Increase with Rutile)		RNA % change (Decrease with Doxy+Cipro)	
	Mean	± SD	Mean	± SD	Mean	± SD	Mean	± SD
AIDS	22.56	2.46	62.70	4.53	23.32	1.74	65.67	4.16
CJD	23.30	1.42	65.07	4.95	23.11	1.52	66.68	3.97
Autism	22.12	2.44	63.69	5.14	23.33	1.35	66.83	3.27
EMF	22.29	2.05	58.70	7.34	22.29	2.05	67.03	5.97
	F value 337.577 P value < 0.001		F value 356.621 P value < 0.001		F value 427.828 P value < 0.001		F value 654.453 P value < 0.001	

Table 3. Effect of rutile and antibiotics on HMG CoA reductase and ATP synthase

Group	HMG CoA R % change (Increase with Rutile)		HMG CoA R % change (Decrease with Doxy+Cipro)		ATP synthase % (Increase with Rutile)		ATP synthase % (Decrease with Doxy+Cipro)	
	Mean	± SD	Mean	± SD	Mean	± SD	Mean	± SD
Normal	4.30	0.20	18.35	0.35	4.40	0.11	18.78	0.11
Schizo	22.91	1.92	61.63	6.79	23.67	1.42	67.39	3.13
Seizure	23.09	1.69	61.62	8.69	23.09	1.90	66.15	4.09
AD	23.43	1.68	61.68	8.32	23.58	2.08	66.21	3.69
MS	23.14	1.85	59.76	4.82	23.52	1.76	67.05	3.00
NHL	22.28	1.76	61.88	6.21	24.01	1.17	66.66	3.84
DM	23.06	1.65	62.25	6.24	23.72	1.73	66.25	3.69
AIDS	22.86	2.58	66.53	5.59	23.15	1.62	66.48	4.17
CJD	22.38	2.38	60.65	5.27	23.00	1.64	66.67	4.21
Autism	22.72	1.89	64.51	5.73	22.60	1.64	66.86	4.21
EMF	22.92	1.48	61.91	7.56	23.37	1.31	63.97	3.62
	F value 319.332 P value < 0.001		F value 199.553 P value < 0.001		F value 449.503 P value < 0.001		F value 673.081 P value < 0.001	

Table 4. Effect of rutile and antibiotics on digoxin and bile acids

Group	Digoxin (ng/ml) (Increase with Rutile)		Digoxin (ng/ml) (Decrease with Doxy+Cipro)		Bile Acids % change (Increase with Rutile)		Bile Acids % change (Decrease with Doxy+Cipro)	
	Mean	± SD	Mean	± SD	Mean	± SD	Mean	± SD
Normal	0.11	0.00	0.054	0.003	4.29	0.18	18.15	0.58
Schizo	0.55	0.06	0.219	0.043	23.20	1.87	57.04	4.27
Seizure	0.51	0.05	0.199	0.027	22.61	2.22	66.62	4.99
AD	0.55	0.03	0.192	0.040	22.12	2.19	62.86	6.28
MS	0.52	0.03	0.214	0.032	21.95	2.11	65.46	5.79
NHL	0.54	0.04	0.210	0.042	22.98	2.19	64.96	5.64
DM	0.47	0.04	0.202	0.025	22.87	2.58	64.51	5.93
AIDS	0.56	0.05	0.220	0.052	22.29	1.47	64.35	5.58
CJD	0.53	0.06	0.212	0.045	23.30	1.88	62.49	7.26
Autism	0.53	0.08	0.205	0.041	22.21	2.04	63.84	6.16
EMF	0.51	0.05	0.213	0.033	23.41	1.41	58.70	7.34
	F value 135.116 P value < 0.001		F value 71.706 P value < 0.001		F value 290.441 P value < 0.001		F value 203.651 P value < 0.001	

Table 5. Effect of rutile and antibiotics on pyruvate and hexokinase

Group	Pyruvate % change (Increase with Rutile)		Pyruvate % change (Decrease with Doxy+Cipro)		Hexokinase % change (Increase with Rutile)		Hexokinase % change (Decrease with Doxy+Cipro)	
	Mean	± SD	Mean	± SD	Mean	± SD	Mean	± SD
Normal	4.34	0.21	18.[illegible]3	0.82	4.21	0.16	18.56	0.76
Schizo	20.99	1.46	61.[illegible]3	9.73	23.01	2.61	65.87	5.27
Seizure	20.94	1.54	62.[illegible]6	8.52	23.33	1.79	62.50	5.56
AD	22.63	0.88	56.[illegible]0	8.59	22.96	2.12	65.11	5.91
MS	21.59	1.23	60.[illegible]8	9.22	22.81	1.91	63.47	5.81
NHL	21.19	1.61	58.[illegible]7	7.47	22.53	2.41	64.29	5.44
DM	20.67	1.38	58.[illegible]5	8.12	23.23	1.88	65.11	5.14

Table 5. (Continued)

Group	Pyruvate % change (Increase with Rutile)		Pyruvate % change (Decrease with Doxy+Cipro)		Hexokinase % change (Increase with Rutile)		Hexokinase % change (Decrease with Doxy+Cipro)	
	Mean	± SD	Mean	± SD	Mean	± SD	Mean	± SD
AIDS	21.21	2.36	58.73	8.10	21.11	2.25	64.20	5.38
CJD	21.07	1.79	63.90	7.13	22.47	2.17	65.97	4.62
Autism	21.91	1.71	58.45	6.66	22.88	1.87	65.45	5.08
EMF	22.29	2.05	62.37	5.05	21.66	1.94	67.03	5.97
	F value 321.255 P value < 0.001		F value 115.242 P value < 0.001		F value 292.065 P value < 0.001		F value 317.966 P value < 0.001	

Table 6. Effect of rutile and antibiotics on hydrogen peroxide and delta amino levulinic acid

Group	H_2O_2 % (Increase with Rutile)		H_2O_2 % (Decrease with Doxy+Cipro)		ALA % (Increase with Rutile)		ALA % (Decrease with Doxy+Cipro)	
	Mean	± SD	Mean	± SD	Mean	± SD	Mean	± SD
Normal	4.43	0.19	18.13	0.63	4.40	0.10	18.48	0.39
Schizo	22.50	1.66	60.21	7.42	22.52	1.90	66.39	4.20
Seizure	23.81	1.19	61.08	7.38	22.83	1.90	67.23	3.45
AD	22.65	2.48	60.19	6.98	23.67	1.68	66.50	3.58
MS	21.14	1.20	60.53	4.70	22.38	1.79	67.10	3.82
NHL	23.35	1.76	59.17	3.33	23.34	1.75	66.80	3.43
DM	23.27	1.53	58.91	6.09	22.87	1.84	66.31	3.68
AIDS	23.32	1.71	63.15	7.62	23.45	1.79	66.32	3.63
CJD	22.86	1.91	63.66	6.88	23.17	1.88	68.53	2.65
Autism	23.52	1.49	63.24	7.36	23.20	1.57	66.65	4.26
EMF	23.29	1.67	60.52	5.38	22.29	2.05	61.91	7.56
	F value 380.721 P value < 0.001		F value 171.228 P value < 0.001		F value 372.716 P value < 0.001		F value 556.411 P value < 0.001	

Table 7. Effect of rutile and antibiotics on dopamine and serotonin

Group	DOPAMINE % (Increase with Rutile)		DOPAMINE % (Decrease with Doxy+Cipro)		5 HT % change (Increase with Rutile)		5 HT % change (Decrease with Doxy+Cipro)	
	Mean	± SD	Mean	± SD	Mean	± SD	Mean	± SD
Normal	4.41	0.15	18.63	0.12	4.34	0.15	18.24	0.37
Schizo	21.88	1.19	56.28	3.60	23.02	1.65	67.61	2.77
Seizure	22.29	1.33	55.38	3.62	22.13	2.14	66.26	3.93
AD	23.66	1.67	55.97	3.36	23.09	1.81	65.86	4.27
MS	22.92	2.14	57.54	3.65	21.93	2.29	63.70	5.63
NHL	23.81	1.90	56.95	3.67	23.12	1.71	65.12	5.58
DM	24.10	1.61	55.78	4.43	22.73	2.46	65.87	4.35
AIDS	23.43	1.57	56.30	3.57	22.98	1.50	65.13	4.87
CJD	23.70	1.75	58.06	3.52	23.81	1.49	64.89	6.01
Autism	22.76	2.20	57.63	3.52	22.79	2.20	64.26	6.02
EMF	22.28	1.52	54.05	2.79	22.82	1.56	64.61	4.95
	F value 403.394 P value < 0.001		F value 680.284 P value < 0.001		F value 348.867 P value < 0.001		F value 364.999 P value < 0.001	

The actinides modulate RNA folding and catalyse its ribozymal action. Digoxin can cut and paste the viroidal strands by modulating RNA splicing generating RNA viroidal diversity. The viroids are evolutionarily escaped archaeal group I introns which have retrotransposition and self splicing qualities.[18] Archaeal pyruvate can produce histone deacetylase inhibition resulting in endogenous retroviral (HERV) reverse transcriptase and integrase expression. This can integrate the RNA viroidal complementary DNA into the noncoding region of eukaryotic non coding DNA using HERV integrase as has been described for borna and ebola viruses.[19] The noncoding DNA is lengthened by integrating RNA viroidal complementary DNA with the integration going on as a continuing event.

The archaea genome can also get integrated into human genome using integrase as has been described for trypanosomes [20]. The integrated viroids and archaea can undergo vertical transmission and can exist as genomic parasites.[19,20] This increases the length and alters the grammar of the noncoding region producing memes or memory of acquired characters as well as eukaryotic speciation and individuality.[21] The viroidal complementary DNA can function as jumping genes producing a dynamic genome important in storage of synaptic information, HLA gene expression and developmental gene expression. The RNA viroids can regulate mrna function by RNA interference.[18] The phenomena of RNA interference can modulate T cell and B cell function, insulin signaling lipid metabolism, cell growth and differentiation, apoptosis, neuronal transmission and euchromatin/ heterochromatin expression.

The archaea and viroids can regulate the nervous system including the NMDA/GABA thalamocorticothalamic pathway mediating conscious perception.[4,22] NMDA/GABA receptors can be modulated by digoxin induced calcium oscillations resulting NMDA/GAD activity induction, PAH increasing NMDA activity and inducing GAD as well as viroid induced RNA interference.[4] The cholesterol ring oxidase generated pyruvate can be converted by the GABA shunt pathway to glutamate and GABA. The dipolar PAH and archaeal magnetite in the setting of digoxin induced sodium potassium ATPase inhibition can produce a pumped phonon system mediated frohlich model superconducting state [22] inducing quantal perception with nanoarchaeal sensed gravity producing the orchestrated reduction of the quantal possibilities to the macrosopic world.[4,22] The archaea can regulate limbic lobe transmission with archaeal cholesterol aromatase/ring oxidase generated norepinephrine, dopamine, serotonin and acetyl choline.[16] The higher degree of integration of the archaea into the genome produces increased digoxin synthesis producing right hemispheric dominance and lesser degree producing left hemispheric dominance.[4] The increased integration of archaea into the neuronal genome can produce increased cholesterol oxidase and aromatase mediated monoamine and NMDA transmission producing schizophrenia and autism. Archaea and RNA viroid can bind the TLR receptor induce NFKB producing immune activation and cytokine TNF alpha secretion. The archaeal DXP and mevalonate pathway metabolites can bind γδ TCR and digoxin induced calcium signaling can activate NFKB producing chronic immune activation.[4,23] The archaea and viroid induced chronic immune activation and generation of superantigens can lead on to autoimmune disease. Archaea, viroids and digoxin can induce the host AKT PI3K, AMPK, HIF alpha and NFKB producing the Warburg metabolic phenotype.[24] The increased glycolytic hexokinase activity, decrease in blood ATP, leakage of cytochrome C, increase in serum pyruvate and decrease in acetyl CoA indicates the generation of the Warburg phenotype. There is induction of glycolysis, inhibition of PDH activity and mitochondrial dysfunction resulting in inefficient energetics and metabolic syndrome. The archaea and viroid generated cytokines can lead to TNF alpha induced insulin resistance and metabolic syndrome x. The accumulated pyruvate enters the gaba shunt pathway and is converted to citrate which is acted upon by citrate lyase and converted to acetyl CoA, used for cholesterol synthesis.[24] The pyruvate can be converted to glutamate and ammonia which is oxidised by archaea for energy needs. The increased cholesterol substrate leads to increased archaeal growth and digoxin synthesis leading to metabolic channeling to the mevalonate pathway. The archaeal bile acids are steroidal hormones which can bind GPCR and modulate D2 regulating the conversion of T4 to T3 which activates uncoupling proteins, can activate NRF ½ inducing NQO1, GST, HOI reducing redox stress, can bind FXR regulating insulin receptor sensitivity and bind PXR inducing the bile acid shunt pathway of cholesterol detoxification [25] The archaea and viroid induced monocyte activation and Warburg phenotype induced increased cholesterol synthesis leads to atherogenesis. The Warburg phenotype induced increased mitochondrial PT pore hexokinase, archaeal PAH and viroid induced RNA interference can lead on to malignant transformation. The digoxin and PAH induced increased intracellular calcium can lead to PT pore dysfunction, cell death and neuronal degeneration [4]

The archaeal cholesterol catabolism can deplete the cell membranes of cholesterol resulting in organelle dysfunction and degeneration. The RNA viroids can recombine with HERV sequences and get encapsulated in microvesicles contributing to the retroviral state. The prion protein conformation is modulated by RNA viroid binding producing prion disease. The archaeal digoxin and rutile induced magnesium depletion can lead MPS deposition and produce EMF, CCP, MNG and mucoid angiopathy [4]

The metal actinides provide radiolytic energy, catalysis for oligomer formation and provide a coordinating ion for metalloenzymes all important in abiogenesis.[7] The metal actinide surfaces would by surface metabolism generate acetate which could get converted to acetyl CoA and then to cholesterol which functions as the primal prebiotic molecule self organizing into self replicating supramolecular systems, the lipid organism [9,26,27] Cholesterol by radiolysis by actinides would have formed PAH generating PAH aromatic organism [9]. Cholesterol radiolysis would generate pyruvate which would get converted to amino acids, sugars, nucleotides, porphyrins, fatty acids and TCA acids. Anastase and rutile surfaces can produce polymerization of amino acids, isoprenyl residues, PAH and nucleotides to generate the initial lipid organism, PAH organism, prions and RNA viroids which would have symbiosed to generate the archaeal protocell. The archaea evolved into gram negative and gram positive bacteria with a mevalonate pathway which had a evolutionary advantage and the symbiosis of archaea with gram negative organism generated the eukaryotic cell [28]. The data supports the persistence of an actinide and cholesterol based shadow biosphere which throws light on the actinide based origin of life and cholesterol as the premier prebiotic molecule.

References

[1] Hanold D., Randies, J.W. (1991). Coconut cadang-cadang disease and its viroid agent, *Plant Disease,* 75, 330-335.

[2] Valiathan M.S., Somers, K., Kartha, C.C. (1993). *Endomyocardial Fibrosis.* Delhi: Oxford University Press.

[3] Edwin B.T., Mohankumaran, C. (2007). Kerala wilt disease phytoplasma: Phylogenetic analysis and identification of a vector, *Proutista moesta, Physiological and Molecular Plant Pathology,* 71(1-3), 41-47.

[4] Kurup R., Kurup, P.A. (2009). *Hypothalamic digoxin, cerebral dominance and brain function in health and diseases.* New York: Nova Science Publishers.

[5] Eckburg P.B., Lepp, P.W., Relman, D.A. (2003). Archaea and their potential role in human disease, *Infect. Immun.,* 71, 591-596.

[6] Smit A., Mushegian, A. (2000). Biosynthesis of isoprenoids via mevalonate in Archaea: the lost pathway, *Genome Res,* 10(10), 1468-84.

[7] Adam Z. (2007). Actinides and Life's Origins, *Astrobiology,* 7, 6-10.

[8] Schoner W. (2002). Endogenous cardiac glycosides, a new class of steroid hormones, *Eur. J. Biochem.,* 269, 2440-2448.

[9] Davies P.C.W., Benner, S.A., Cleland, C.E., Lineweaver, C.H., McKay, C.P., Wolfe-Simon, F. (2009). Signatures of a Shadow Biosphere, *Astrobiology,* 10, 241-249.

[10] Richmond W. (1973). Preparation and properties of a cholesterol oxidase from nocardia species and its application to the enzymatic assay of total cholesterol in serum, *Clin. Chem.,* 19, 1350-1356.

[11] Snell E.D., Snell, C.T. (1961). *Colorimetric Methods of Analysis.* Vol 3A. New York: Van NoStrand.

[12] Glick D. (1971). *Methods of Biochemical Analysis.* Vol 5. New York: Interscience Publishers.

[13] Colowick, Kaplan, N.O. (1955). *Methods in Enzymology.* Vol 2. New York: Academic Press.

[14] Van der Geize R., Yam, K., Heuser, T., Wilbrink, M.H., Hara, H., Anderton, M.C. (2007). A gene cluster encoding cholesterol catabolism in a soil actinomycete provides insight into Mycobacterium tuberculosis survival in macrophages, *Proc. Natl. Acad. Sci. USA,* 104(6), 1947-52.

[15] Francis A.J. (1998). Biotransformation of uranium and other actinides in radioactive wastes, *Journal of Alloys and Compounds,* 271(273), 78-84.

[16] Probian C., Wülfing, A., Harder, J. (2003). Anaerobic mineralization of quaternary carbon atoms: Isolation of denitrifying bacteria on pivalic acid (2,2-Dimethylpropionic acid), *Applied and Environmental Microbiology,* 69(3), 1866-1870.

[17] Vainshtein M., Suzina, N., Kudryashova, E., Ariskina, E. (2002). New Magnet-Sensitive Structures in Bacterial and Archaeal Cells, *Biol. Cell,* 94(1), 29-35.

[18] Tsagris E.M., de Alba, A.E., Gozmanova, M., Kalantidis, K. (2008). Viroids, *Cell Microbiol.,* 10, 2168.

[19] Horie M., Honda, T., Suzuki, Y., Kobayashi, Y., Daito, T., Oshida, T. (2010). Endogenous non-retroviral RNA virus elements in mammalian genomes, *Nature,* 463, 84-87.

[20] Hecht M., Nitz, N., Araujo, P., Sousa, A., Rosa, A., Gomes, D. (2010). Genes from Chagas parasite can transfer to humans and be passed on to children. Inheritance of DNA Transferred from American Trypanosomes to Human Hosts, *PLoS ONE,* 5, 2-10.

[21] Flam F. (1994). Hints of a language in junk DNA, *Science,* 266, 1320.

[22] Lockwood, M. (1989). *Mind, Brain and the Quantum.* Oxford: B. Blackwell.

[23] Eberl M., Hintz, M., Reichenberg, A., Kollas, A., Wiesner, J., Jomaa, H. (2010). Microbial isoprenoid biosynthesis and human γδ T cell activation, *FEBS Letters,* 544(1), 4-10.

[24] Wallace D.C. (2005). Mitochondria and Cancer: Warburg Addressed, *Cold Spring Harbor Symposia on Quantitative Biology,* 70, 363-374.

[25] Lefebvre P., Cariou, B., Lien, F., Kuipers, F., Staels, B. (2009). Role of Bile Acids and Bile Acid Receptors in Metabolic Regulation, *Physiol. Rev.,* 89(1), 147 - 191.

[26] Wächtershäuser, G. (1988). Before enzymes and templates: theory of surface metabolism. *Microbiol. Rev.*, 52(4), 452-84.

[27] Russell, M.J., Martin W. (2004). The rocky roots of the acetyl-CoA Pathway. *Trends in Biochemical Sciences,* 29(7).

[28] Margulis, L. (1996). Archaeal-eubacterial mergers in the origin of Eukarya: phylogenetic classification of life. *Proc. Natl. Acad. Sci. USA,* 93, 1071-1076.

Chapter II

A Cholesterol and Actinide Dependent Shadow Biosphere of Archaea and Viroids Indicating Cholesterol Based Abiogenesis- Evolution of the Biological Universe

Introduction

Endomyocardial fibrosis (EMF) along with the root wilt disease of coconut is endemic to Kerala with its radioactive actinide beach sands. Actinides like rutile producing intracellular magnesium deficiency due to rutile-magnesium exchange sites in the cell membrane has been implicated in the etiology of EMF [1]. Endogenous digoxin, a steroidal glycoside which functions as a membrane sodium-potassium ATPase inhibitor has also been related to its etiology due to the intracellular magnesium deficiency it produces [2]. Organisms like phytoplasmas and viroids have also been demonstrated to play a role in the etiology of these diseases [3,4]. Endogenous digoxin has been related to the pathogenesis of schizophrenia, malignancy, metabolic syndrome x, autoimmune disease and neuronal degeneration [2]. The possibility of endogenous digoxin synthesis by actinide based primitive organism like archaea with a mevalonate pathway and cholesterol catabolism was considered [5,6,7]. Davies has put forward the concept of a shadow biosphere of organisms with alternate biochemistry present in earth itself [8]. An actinide dependent shadow biosphere of archaea and viroids in the above mentioned disease states is described [6]. Metal actinides in beach sands have been postulated to play a role in abiogenesis [6]. Actinide mineral like rutile, monazite and illmenite by surface metabolism would have contributed to abiogenesis [9]. A hypothesis of cholesterol as the primal prebiotic molecule synthesised on actinide surfaces with all other biomolecules arising from it and a self replicating cholesterol lipid organism as the initial life form is presented. The role of actinidic archaea in the genesis of the interstellar polycyclic aromatic hydrocarbons as well as the interstellar magnetic fields important in the evolution of the universe is hypothesized.

Materials and Methods

Informed consent of the subjects and the approval of the ethics committee were obtained for the study. The following groups were included in the study:- endomyocardial fibrosis, alzheimer's disease, multiple sclerosis, non-hodgkin's lymphoma, metabolic syndrome x with cerebrovascular thrombosis and coronary artery disease, schizophrenia, autism, seizure disorder, creutzfeldt jakob disease and acquired immunodeficiency syndrome. There were 10 patients in each group and each patient had an age and sex matched healthy control selected randomly from the general population. The blood samples were drawn in the fasting state before treatment was initiated. Plasma from fasting heparinised blood was used and the experimental protocol was as follows (I) Plasma+phosphate buffered saline, (II) same as I+cholesterol substrate, (III) same as II+rutile 0.1 mg/ml, (IV) same as II+ciprofloxacine and doxycycline each in a concentration of 1 mg/ml. Cholesterol substrate was prepared as described by Richmond [10]. Aliquots were withdrawn at zero time immediately after mixing and after incubation at 37°C for 1 hour. The following estimations were carried out:- Cytochrome F420, free RNA, free DNA, muramic acid, polycyclic aromatic hydrocarbon, hydrogen peroxide, serotonin, pyruvate, ammonia, glutamate, cytochrome C, hexokinase, ATP synthase, HMG CoA redutase, digoxin and bile acids [11,12,13,14]. Cytochrome F420 was estimated flourimetrically (excitation wavelength 420 nm and emission wavelength 520 nm). Polycyclic aromatic hydrocarbon was estimated by measuring hydrogen peroxide liberated by using glucose reagent. The statistical analysis was done by ANOVA.

Results

The parameters checked as indicated above were:- cytochrome F420, free RNA, free DNA, muramic acid, polycyclic aromatic hydrocarbon, hydrogen peroxide, serotonin, pyruvate, ammonia, glutamate, cytochrome C, hexokinase, ATP synthase, HMG CoA redutase, digoxin and bile acids. Plasma of control subjects showed increased levels of the above mentioned parameters with after incubation for 1 hour and addition of cholesterol substrate resulted in still further significant increase in these parameters. The plasma of patients showed similar results but the extent of increase was more. The addition of antibiotics to the control plasma casued a decrease in all the parameters while addition of rutile increased their levels. The addition of antibiotics to the patient's plasma caused a decrease in all the parameters while addition of rutile increased their levels but the extent of change was more in patient's sera as compared to controls. The results are expressed in tables 1-7 as percentage change in the parameters after 1 hour incubation as compared to the values at zero time.

Discussion

There was increase in cytochrome F420 indicating archaeal growth. The archaea can synthesise and use cholesterol as a carbon and energy source [15,16]. The archeal origin of the enzyme activities was indicated by antibiotic induced suppression. The study indicates the presence of actinide based archaea with an alternate actinide based enzymes or metalloenzymes in the system as indicated by rutile induced increase in enzyme activities [17]. There was also an increase in archaeal HMG CoA reductase activity indicating increased cholesterol synthesis by the archaeal mevalonate pathway. The archaeal beta hydroxyl steroid dehydrogenase activity indicating digoxin synthesis and archaeal cholesterol hydroxylase activity indicating bile acid synthesis were increased [7].

The archaeal cholesterol oxidase activity was increased resulting in generation of pyruvate and hydrogen peroxide [16]. The pyruvate gets converted to glutamate and ammonia by the GABA shunt pathway.

The archaeal aromatization of cholesterol generating PAH, serotonin and dopamine was also detected [18]. The archaeal glycolytic hexokinase activity and archaeal extracellular ATP synthase activity were increased. The archaea can undergo magnetite and calcium carbonate mineralization and can exist as calcified nanoforms [19]. There was an increase in free RNA indicating self replicating RNA viroids and free DNA indicating generation of viroid complementary DNA strands by archaeal reverse transcriptase activity. The actinides modulate RNA folding and catalyse its ribozymal action. Digoxin can cut and paste the viroidal strands by modulating RNA splicing generating RNA viroidal diversity. The viroids are evolutionarily escaped archaeal group I introns which have retrotransposition and self splicing qualities [20]. Archaeal pyruvate can produce histone deacetylase inhibition resulting in endogenous retroviral (HERV) reverse transcriptase and integrase expression. This can integrate the RNA viroidal complementary DNA into the noncoding region of eukaryotic non coding DNA using HERV integrase as has been described for borna and ebola viruses [21]. The noncoding DNA is lengthened by integrating RNA viroidal complementary DNA with the integration going on as a continuing event. The archaea genome can also get integrated into human genome using integrase as has been described for trypanosomes [22]. The integrated viroids and archaea can undergo vertical transmission and can exist as genomic parasites [21,22].

This increases the length and alters the grammar of the noncoding region producing memes or memory of acquired characters [23]. The viroidal complementary DNA can function as jumping genes producing a dynamic genome important in storage of synaptic information, HLA gene expression and developmental gene expression. The RNA viroids can regulate mrna function by RNA interference [20]. The phenomena of RNA interference can modulate T cell and B cell function, insulin signaling lipid metabolism, cell growth and differentiation, apoptosis, neuronal transmission and euchromatin/heterochromatin expression.

The presence of muramic acid, HMG CoA reductase and cholesterol oxidase activity inhibited by antibiotics indicates the presence of bacteria with mevalonate pathway. The bacterial with mevalonate pathway include streptococcus, staphylococcus, actinomycetes, listeria, coxiella and borrelia [24]. The bacteria and archaea with mevalonate pathway and cholesterol catabolism had a evolutionarily advantage and constitutes the isoprenoidal clade organism with the archaea evolving into mevalonate pathway gram positive and gram negative organism through horizontal gene transfer of viroidal and virus genes [25]. The isoprenoidal clade prokaryotes develop into other groups of prokaryotes via viroidal/virus as well as eukaryotic horizontal gene transfer producing bacterial speciation [26].

Table 1. Effect of rutile and antibiotics on cytochrome F420 and muramic acid

Group	CYT F420 % (Increase with Rutile)		CYT F420 % (Decrease with Doxy+Cipro)		Muramic acid % change (Increase with Rutile)		Muramic acid % change (Decrease with Doxy+Cipro)	
	Mean	± SD	Mean	± SD	Mean	± SD	Mean	± SD
Normal	4.48	0.15	18.24	0.66	4.45	0.14	18.25	0.72
Schizo	23.24	2.01	58.72	7.08	23.01	1.69	59.49	4.30
Seizure	23.46	1.87	59.27	8.86	22.67	2.29	57.69	5.29
AD	23.12	2.00	56.90	6.94	23.26	1.53	60.91	7.59
MS	22.12	1.81	61.33	9.82	22.83	1.78	59.84	7.62
NHL	22.79	2.13	55.90	7.29	22.84	1.42	66.07	3.78
DM	22.59	1.86	57.05	8.45	23.40	1.55	65.77	5.27
AIDS	22.29	1.66	59.02	7.50	23.23	1.97	65.89	5.05
CJD	22.06	1.61	57.81	6.04	23.46	1.91	61.56	4.61
Autism	21.68	1.90	57.93	9.64	22.61	1.42	64.48	6.90
EMF	22.70	1.87	60.46	8.06	23.73	1.38	65.20	6.20
	F value 306.749 P value < 0.001		F value 130.054 P value < 0.001		F value 391.318 P value < 0.001		F value 257.996 P value < 0.001	

Table 2. Effect of rutile and antibiotics on free RNA and DNA

Group	DNA % change (Increase with Rutile)		DNA % change (Decrease with Doxy+Cipro)		RNA % change (Increase with Rutile)		RNA % change (Decrease with Doxy+Cipro)	
	Mean	± SD	Mean	± SD	Mean	± SD	Mean	± SD
Normal	4.37	0.15	18.39	0.38	4.37	0.13	18.38	0.48
Schizo	23.28	1.70	61.41	3.36	23.59	1.83	65.69	3.94
Seizure	23.40	1.51	63.68	4.66	23.08	1.87	65.09	3.48
AD	23.52	1.65	64.15	4.60	23.29	1.92	65.39	3.95
MS	22.62	1.38	63.82	5.53	23.29	1.98	67.46	3.96
NHL	22.42	1.99	61.14	3.47	23.78	1.20	66.90	4.10
DM	23.01	1.67	65.35	3.56	23.33	1.86	66.46	3.65

Group	DNA % change (Increase with Rutile)		DNA % change (Decrease with Doxy+Cipro)		RNA % change (Increase with Rutile)		RNA % change (Decrease with Doxy+Cipro)	
	Mean	± SD	Mean	± SD	Mean	± SD	Mean	± SD
AIDS	22.56	2.46	62.[illegible]0	4.53	23.32	1.74	65.67	4.16
CJD	23.30	1.42	65.[illegible]7	4.95	23.11	1.52	66.68	3.97
Autism	22.12	2.44	63.[illegible]9	5.14	23.33	1.35	66.83	3.27
EMF	22.29	2.05	58.[illegible]0	7.34	22.29	2.05	67.03	5.97
	F value 337.577 P value < 0.001		F value 356.621 P value < 0.001		F value 427.828 P value < 0.001		F value 654.453 P value < 0.001	

Table 3. Effect of rutile and antibiotics on HMG CoA reductase and ATP synthase

Group	HMG CoA R % change (Increase with Rutile)		HMG CoA R % change (Decrease with Doxy+Cipro)		ATP synthase % (Increase with Rutile)		ATP synthase % (Decrease with Doxy+Cipro)	
	Mean	± SD	Mean	± SD	Mean	± SD	Mean	± SD
Normal	4.30	0.20	18.35	0.35	4.40	0.11	18.78	0.11
Schizo	22.91	1.92	61.63	6.79	23.67	1.42	67.39	3.13
Seizure	23.09	1.69	61.62	8.69	23.09	1.90	66.15	4.09
AD	23.43	1.68	61.68	8.32	23.58	2.08	66.21	3.69
MS	23.14	1.85	59.76	4.82	23.52	1.76	67.05	3.00
NHL	22.28	1.76	61.88	6.21	24.01	1.17	66.66	3.84
DM	23.06	1.65	62.25	6.24	23.72	1.73	66.25	3.69
AIDS	22.86	2.58	66.53	5.59	23.15	1.62	66.48	4.17
CJD	22.38	2.38	60.65	5.27	23.00	1.64	66.67	4.21
Autism	22.72	1.89	64.51	5.73	22.60	1.64	66.86	4.21
EMF	22.92	1.48	61.91	7.56	23.37	1.31	63.97	3.62
	F value 319.332 P value < 0.001		F value 199.553 P value < 0.001		F value 449.503 P value < 0.001		F value 673.081 P value < 0.001	

Table 4. Effect of rutile and antibiotics on digoxin and bile acids

Group	Digoxin (ng/ml) (Increase with Rutile)		Digoxin (ng/ml) (Decrease with Doxy+Cipro)		Bile Acids % change (Increase with Rutile)		Bile Acids % change (Decrease with Doxy+Cipro)	
	Mean	± SD	Mean	± SD	Mean	± SD	Mean	± SD
Normal	0.11	0.00	0.054	0.003	4.29	0.18	18.15	0.58
Schizo	0.55	0.06	0.219	0.043	23.20	1.87	57.04	4.27
Seizure	0.51	0.05	0.199	0.027	22.61	2.22	66.62	4.99
AD	0.55	0.03	0.192	0.040	22.12	2.19	62.86	6.28
MS	0.52	0.03	0.214	0.032	21.95	2.11	65.46	5.79
NHL	0.54	0.04	0.210	0.042	22.98	2.19	64.96	5.64
DM	0.47	0.04	0.202	0.025	22.87	2.58	64.51	5.93
AIDS	0.56	0.05	0.220	0.052	22.29	1.47	64.35	5.58
CJD	0.53	0.06	0.212	0.045	23.30	1.88	62.49	7.26
Autism	0.53	0.08	0.205	0.041	22.21	2.04	63.84	6.16
EMF	0.51	0.05	0.213	0.033	23.41	1.41	58.70	7.34
	F value 135.116 P value < 0.001		F value 71.706 P value < 0.001		F value 290.441 P value < 0.001		F value 203.651 P value < 0.001	

Table 5. Effect of rutile and antibiotics on pyruvate and hexokinase

Group	Pyruvate % change (Increase with Rutile)		Pyruvate % change (Decrease with Doxy+Cipro)		Hexokinase % change (Increase with Rutile)		Hexokinase % change (Decrease with Doxy+Cipro)	
	Mean	± SD	Mean	± SD	Mean	± SD	Mean	± SD
Normal	4.34	0.21	18.43	0.82	4.21	0.16	18.56	0.76
Schizo	20.99	1.46	61.23	9.73	23.01	2.61	65.87	5.27
Seizure	20.94	1.54	62.76	8.52	23.33	1.79	62.50	5.56
AD	22.63	0.88	56.40	8.59	22.96	2.12	65.11	5.91
MS	21.59	1.23	60.28	9.22	22.81	1.91	63.47	5.81
NHL	21.19	1.61	58.57	7.47	22.53	2.41	64.29	5.44
DM	20.67	1.38	58.75	8.12	23.23	1.88	65.11	5.14
AIDS	21.21	2.36	58.73	8.10	21.11	2.25	64.20	5.38

Table 5. (Continued)

Group	Pyruvate % change (Increase with Rutile)		Pyruvate % change (Decrease with Doxy+Cipro)		Hexokinase % change (Increase with Rutile)		Hexokinase % change (Decrease with Doxy+Cipro)	
	Mean	± SD	Mean	± SD	Mean	± SD	Mean	± SD
CJD	21.07	1.79	63.[illegible]	7.13	22.47	2.17	65.97	4.62
Autism	21.91	1.71	58.45	6.66	22.88	1.87	65.45	5.08
EMF	22.29	2.05	62.[illegible]	5.05	21.66	1.94	67.03	5.97
	F value 321.255 P value < 0.001		F value 115.242 P value < 0.001		F value 292.065 P value < 0.001		F value 317.966 P value < 0.001	

Table 6. Effect of rutile and antibiotics on hydrogen peroxide and delta amino levulinic acid

Group	H_2O_2 % (Increase with Rutile)		H_2O_2 % (Decrease with Doxy+Cipro)		ALA % (Increase with Rutile)		ALA % (Decrease with Doxy+Cipro)	
	Mean	± SD	Mean	± SD	Mean	± SD	Mean	± SD
Normal	4.43	0.19	18.1[illegible]	0.63	4.40	0.10	18.48	0.39
Schizo	22.50	1.66	60.2[illegible]	7.42	22.52	1.90	66.39	4.20
Seizure	23.81	1.19	61.0[illegible]	7.38	22.83	1.90	67.23	3.45
AD	22.65	2.48	60.1[illegible]	6.98	23.67	1.68	66.50	3.58
MS	21.14	1.20	60.5[illegible]	4.70	22.38	1.79	67.10	3.82
NHL	23.35	1.76	59.1[illegible]	3.33	23.34	1.75	66.80	3.43
DM	23.27	1.53	58.9[illegible]	6.09	22.87	1.84	66.31	3.68
AIDS	23.32	1.71	63.1[illegible]	7.62	23.45	1.79	66.32	3.63
CJD	22.86	1.91	63.6[illegible]	6.88	23.17	1.88	68.53	2.65
Autism	23.52	1.49	63.2[illegible]	7.36	23.20	1.57	66.65	4.26
EMF	23.29	1.67	60.5[illegible]	5.38	22.29	2.05	61.91	7.56
	F value 380.721 P value < 0.001		F value 171.228 P value < 0.001		F value 372.716 P value < 0.001		F value 556.411 P value < 0.001	

Table 7. Effect of rutile and antibiotics on PAH and serotonin

Group	PAH % (Increase with Rutile)		PAH % (Decrease with Doxy+Cipro)		5 HT % change (Increase with Rutile)		5 HT % change (Decrease with Doxy+Cipro)	
	Mean	± SD	Mean	± SD	Mean	± SD	Mean	± SD
Normal	4.41	0.15	18.63	0.12	4.34	0.15	18.24	0.37
Schizo	21.88	1.19	66.28	3.60	23.02	1.65	67.61	2.77
Seizure	22.29	1.33	65.38	3.62	22.13	2.14	66.26	3.93
AD	23.66	1.67	65.97	3.36	23.09	1.81	65.86	4.27
MS	22.92	2.14	67.54	3.65	21.93	2.29	63.70	5.63
NHL	23.81	1.90	66.95	3.67	23.12	1.71	65.12	5.58
DM	24.10	1.61	65.78	4.43	22.73	2.46	65.87	4.35
AIDS	23.43	1.57	66.30	3.57	22.98	1.50	65.13	4.87
CJD	23.70	1.75	68.06	3.52	23.81	1.49	64.89	6.01
Autism	22.76	2.20	67.63	3.52	22.79	2.20	64.26	6.02
EMF	22.28	1.52	64.05	2.79	22.82	1.56	64.61	4.95
	F value 403.394 P value < 0.001		F value 680.284 P value < 0.001		F value 348.867 P value < 0.001		F value 364.999 P value < 0.001	

The RNA viroids and its complementary DNA developed into cholesterol enveloped RNA and DNA viruses like herpes, retrovirus, influenza virus, borna virus, cytomegalo virus and Ebstein Barr virus by recombining with eukaryotic and human genes resulting in viral speciation. Bacterial and viral species are ill defined and fuzzy with all of them forming one common genetic pool with frequent horizontal gene transfer and recombination. Thus the multi and unicellular eukaryote with its genes serves the purpose of prokaryotic and viral speciation .The multicellular eukaryote developed so that their endosymbiotic archaeal colonies could survive and forage better .The multicellular eukaryotes are like bacterial biofilms. The archaea and bacteria with a mevalonate pathway uses the extracellular RNA viroids and DNA viroids for quorum sensing and in the generation of symbiotic biofilm like structures which develop into multicellular eukaryotes [27,28]. The endosymbiotic archaea and bacteria with mevalonate pathway still uses the RNA viroids and DNA viroids for the regulation of muticellular eukaryote. Pollution is induced by the primitive nanoarchaea and mevalonate pathway bacteria synthesised PAH and methane leading on to redox stress. Redox stress leads to sodium potassium ATPase inhibition, inward movement of plasma membrane cholesterol, defective SREBP sensing, increased cholesterol synthesis and nanoarchaeal/mevalonate pathway bacterial growth [29]. Redox stress leads on to viroidal and archaeal multiplication. Redox stress can also lead to HERV reverse transcriptase and integrase expression. The noncoding DNA is formed of integrating RNA viroidal complementary DNA and archaea with the integration going on as a continuing event. The archaeal pox like dsDNA virus forms evolutionarily the nucleus. The integrated viroidal, archaeal and mevalonate pathway bacterial sequences can undergo vertical transmission and can exist as genomic parasites. The genomic integrated archaea, mevalonate pathway bacteria and viroids form a genomic reserve of bacteria and viruses which can recombine with human and eukaryotic genes producing bacterial and viral speciation. The change in the length and grammar of the noncoding region produces eukaryotic speciation and individuality [30]. The integration of nanoarchaea, mevalonate pathway prokaryotes and viroids in to the eukaryotic and human genome produces a chimera which can multiply producing biofilm like multicellular structures having a mixed archaeal, viroidal, prokaryotic and eukaryotic characters which is a regression from the multicellular eukaryotic tissue. This results in a new neuronal, metabolic, immune and tissue phenotype leading to human disease.

The archaea and viroids can regulate the nervous system including the NMDA/GABA thalamocorticothalamic pathway mediating conscious perception [2,31]. NMDA/GABA receptors can be modulated by digoxin induced calcium oscillations resulting NMDA/GAD activity induction, PAH increasing NMDA activity and inducing GAD as well as viroid induced RNA interference [2]. The cholesterol ring oxidase generated pyruvate can be converted by the GABA shunt pathway to glutamate and GABA. The dipolar PAH and archaeal magnetite in the setting of digoxin induced sodium potassium ATPase inhibition can produce a pumped phonon system mediated frohlich model superconducting state inducing quantal perception with nanoarchaeal sensed gravity producing the orchestrated reduction of the quantal possibilities to the macrosopic world [2,31]. The archaea can regulate limbic lobe transmission with archaeal cholesterol aromatase/ring oxidase generated norepinephrine, dopamine, serotonin and acetyl choline [18]. The higher degree of integration of the archaea into the genome produces increased digoxin synthesis producing right hemispheric dominance and lesser degree producing left hemispheric dominance [2]. The increased integration of archaea into the neuronal genome can produce increased cholesterol oxidase and aromatase mediated monoamine and NMDA transmission producing schizophrenia and autism. Archaea and RNA viroid can bind the TLR receptor induce NFKB producing immune activation and cytokine TNF alpha secretion. The archaeal DXP and mevalonate pathway metabolites can bind γδ TCR and digoxin induced calcium signaling can activate NFKB producing chronic immune activation [2,32]. The archaea and viroid induced chronic immune activation and generation of superantigens can lead on to autoimmune disease. Archaea, viroids and digoxin can induce the host AKT PI3K, AMPK, HIF alpha and NFKB producing the Warburg metabolic phenotype [33]. The increased glycolytic hexokinase activity, decrease in blood ATP, leakage of cytochrome C, increase in serum pyruvate and decrease in acetyl CoA indicates the generation of the Warburg phenotype. There is induction of glycolysis, inhibition of PDH activity and mitochondrial dysfunction resulting in inefficient energetics and metabolic syndrome. The archaea and viroid generated cytokines can lead to TNF alpha induced insulin resistance and metabolic syndrome x. The accumulated pyruvate enters the GABA shunt pathway and is converted to citrate which is acted upon by citrate lyase and converted to acetyl CoA, used for cholesterol synthesis [33]. The pyruvate can be converted to glutamate and ammonia which is oxidised by archaea for energy needs. The increased cholesterol substrate leads to increased archaeal growth and digoxin synthesis leading to metabolic channeling to the mevalonate pathway. The archaeal bile acids are steroidal hormones which can bind GPCR and modulate D2 regulating the conversion of T4 to T3 which activates uncoupling proteins, can activate NRF ½ inducing NQO1, GST, HOI reducing redox stress, can bind FXR regulating insulin receptor sensitivity and bind PXR inducing the bile acid shunt pathway of cholesterol detoxification[34]. The archaea and viroid induced monocyte activation and Warburg phenotype induced increased cholesterol synthesis leads to atherogenesis. The Warburg phenotype induced increased mitochondrial PT pore hexokinase, archaeal PAH and viroid induced RNA interference can lead on to malignant transformation. The digoxin and PAH induced increased intracellular calcium can lead to PT pore dysfunction, cell death and neuronal degeneration [2]. The archaeal cholesterol catabolism can deplete the cell membranes of cholesterol resulting in organelle dysfunction and degeneration. The RNA viroids can recombine with HERV sequences and get encapsulated in microvesicles contributing to the retroviral state. The prion protein conformation is modulated by RNA viroid binding producing prion disease. The archaeal digoxin and rutile induced magnesium depletion can lead MPS deposition and produce EMF, CCP, MNG and mucoid angiopathy [2].

The metal actinides provide radiolytic energy, catalysis for oligomer formation and provide a coordinating ion for metalloenzymes all important in abiogenesis [6]. The metal actinide surfaces would by surface metabolism generate acetate which could get converted to acetyl CoA and then to cholesterol which functions as the primal prebiotic molecule self organizing into self replicating supramolecular systems, the lipid organism [8,9,35]. Cholesterol by radiolysis by actinides would have formed PAH generating PAH aromatic organism [8]. Cholesterol radiolysis would generate pyruvate which would get converted to amino acids, sugars, nucleotides, porphyrins, fatty acids and TCA acids. Anastase and rutile surfaces can produce polymerization of amino acids, isoprenyl residues, PAH and nucleotides to generate the initial lipid organism, PAH organism, prions and RNA viroids which would have symbiosed to generate the archaeal protocell. The archaea evolved into gram negative and gram positive bacteria with a mevalonate pathway which had a evolutionary advantage and the symbiosis of archaea with gram negative organism generated the eukaryotic cell [36].

The data supports the persistence of an actinide and cholesterol based shadow biosphere which throws light on the actinide based origin of life and cholesterol as the premier prebiotic molecule.

The archaea can synthesise magnetite by biomineralization. The archaeal cholesterol catabolism can generate PAH. The archaea can exist as nanoarchaea and can have calcified nano forms. The actinidic magnetotactic nanoarchaea and its secreted PAH organisms are extremophiles and survive in the interstellar space and can contribute to the interstellar grains and magnetic fields which play a role in the formation of the galaxies and star systems [37]. The cosmic dust grains occupy the intergalactic space and are thought to be formed of magnetotactic bacteria identified according to their spectral signatures. According to the Hoyle's hypothesis, the cosmic dust magnetotactic bacteria plays a role in the formation of the intergalactic magnetic field. A magnetic field equal in strength to about one millionth part of the magnetic field of earth exists throughout much of our galaxy. The magnetic files can be used to trace the spiral arms of the galaxy following a pattern of field lines that connect young stars and dust in which new stars are formed at a rapid rate. Studies have shown that a fraction of the dust particles have elongated shape similar to bacilli and they are systematically lined up in our galaxy. Moreover the direction of alignment is such that the long axes of the dust tend to be at right angles to the direction of the galactic magnetic field at every point. Magnetotactic bacteria have the property to affect the degree of alignment that is observed. The fact that the magnetotactic bacteria appear to be connected to the magnetic field lines that thread through the spiral arms of the galaxy connecting one region of star formation to another support a role for them in star formation and in the mass distribution and rotation of stars. The nutrient supply for a population of interstellar bacteria comes from mass flows out of supernovas populating the galaxy.

Giants arising in the evolution of such stars experience a phenomenon in which material containing nitrogen, carbon monoxide, hydrogen, helium, water and trace elements essential for life flows continuously outward into space. The interstellar bacteria need liquid water. Water exists only as vapour or solid in the interstellar space and only through star formation leading to associated planets and cometary bodies can there be access to liquid water. To control conditions leading to star formation is of paramount importance in cosmic biology. The rate of star formation is controlled by two factors: Too high a rate of star formation produces a destructive effect of UV radiation and destroys cosmic biology. Star formation as stated before produces water crucial for bacterial growth. Cosmic biology of magnetotactic bacteria and star formation are thus closely interlinked. Systems like solar systems do not arise in random condensation of blobs of interstellar gas.

Only by a rigorous control of rotation of various parts of the system would galaxies and solar system evolved. The key to maintaining control over rotation seems to lie in the intergalactic magnetic field as indeed the whole phenomena of star formation. The intergalactic magnetic fields owes its origin to the lining up of magnetotactic bacteria and the cosmic biology of interstellar bacteria can prosper only by maintaining a firm grip on the interstellar magnetic field and hence on the rate of star formation and type of star system produced. This points to a cosmic intelligence or brain capable of computation, analysis and exploration of the universe at large- of magnetotactic bacterial networks. The origin of life on earth according to the Hoyle's hypothesis would be by seeding of bacteria from the outer intergalactic space.

Comets carrying microorganisms would have interacted with the earth. A thin skin of graphitized material around a single bacteria or clumps of bacteria can shield the interior from destruction by UV light. The sudden surge and diversification of species of plants and animals and their equally sudden extinction has seen from fossil records point to sporadic evolution produced by induction of fresh cometary genes with the arrival of each major new crop of comets [38,39]. The interstellar PAH aromatic organism is formed from nanoarchaeal cholesterol catabolism. The PAH and cholesterol are the interconvertable primal prebiotic molecules. PAH aromatic organism and nanoarchaeal magnetite can have a wave particle existence and bridge the world of bosons and fermions. The nanoarchaea can form biofilms and the PAH aromatic organism can form a molecular quantum computing cloud in the biofilm which forms a interstellar intelligence regulating the formation of star systems and galaxies.The magnetite loaded nanoarchaeal biofilms and PAH aromatic organism quantal computing cloud can bridge the wave particle world functioning as the anthropic observer sensing gravity which orchestrates the reduction of the quantal world of possibilities in to the macroscopic world. The actinide based nanoarchaea can regulate the earth's carbon cycle by methanogenesis, nitrogen cycle by ammonia oxidation and rain formation by contributing the seeding nucleus. The earth's temperature and global warming and cooling are regulated by nanoarchaeal synthesised PAH from cholesterol and methanogenesis. The increased nanoarchaeal growth in ocean beds and soil leads to increased methane production and movement of the earth's crust producing tsunamis and massive earthquake leading to catastrophic mass extinction [40]. The eternal nanoarchaea survive and start the cycle of evolution once more. The actinide based nanoarchaea regulates the human system and biological universe.

References

[1] Valiathan M.S., Somers, K., Kartha, C.C. (1993). *Endomyocardial Fibrosis*. Delhi: Oxford University Press.

[2] Kurup R., Kurup, P.A. (2009). *Hypothalamic digoxin, cerebral dominance and brain function in health and diseases*. New York: Nova Science Publishers.

[3] Hanold D., Randies, J.W. (1991). Coconut cadang-cadang disease and its viroid agent, *Plant Disease,* 75, 330-335.

[4] Edwin B.T., Mohankumaran, C. (2007). Kerala wilt disease phytoplasma: Phylogenetic analysis and identification of a vector, *Proutista moesta, Physiological and Molecular Plant Pathology,* 71(1-3), 41-47.

[5] Eckburg P.B., Lepp, P.W., Relman, D.A. (2003). Archaea and their potential role in human disease, *Infect Immun,* 71, 591-596.

[6] Adam Z. (2007). Actinides and Life's Origins, *Astrobiology,* 7, 6-10.

[7] Schoner W. (2002). Endogenous cardiac glycosides, a new class of steroid hormones, *Eur. J. Biochem.,* 269, 2440-2448.

[8] Davies P.C.W., Benner, S.A., Cleland, C.E., Lineweaver, C.H., McKay, C.P., Wolfe-Simon, F. (2009). Signatures of a Shadow Biosphere, *Astrobiology,* 10, 241-249.

[9] Wächtershäuser G. (1988). Before enzymes and templates: theory of surface metabolism, *Microbiol. Rev.,* 52(4), 452-84.

[10] Richmond W. (1973). Preparation and properties of a cholesterol oxidase from nocardia species and its application to the enzymatic assay of total cholesterol in serum, *Clin. Chem.,* 19, 1350-1356.

[11] Snell E.D., Snell, C.T. (1961). *Colorimetric Methods of Analysis.* Vol 3A. New York: Van NoStrand.

[12] Glick D. (1971). *Methods of Biochemical Analysis.* Vol 5. New York: Interscience Publishers.

[13] Colowick, Kaplan, N.O. (1955). *Methods in Enzymology.* Vol 2. New York: Academic Press.

[14] Maarten A.H., Marie-Jose, M., Cornelia, G., van Helden–Meewsen, Fritz, E., Marten, P.H. (1995). Detection of muramic acid in human spleen, *Infection and Immunity,* 63(5), 1652 – 1657.

[15] Smit A., Mushegian, A. (2000). Biosynthesis of isoprenoids via mevalonate in Archaea: the lost pathway, *Genome Res,* 10(10), 1468-84.

[16] Van der Geize R., Yam, K., Heuser, T., Wilbrink, M.H., Hara, H., Anderton, M.C. (2007). A gene cluster encoding cholesterol catabolism in a soil actinomycete provides insight into Mycobacterium tuberculosis survival in macrophages, *Proc. Natl. Acad. Sci. USA,* 104(6), 1947-52.

[17] Francis A.J. (1998). Biotransformation of uranium and other actinides in radioactive wastes, *Journal of Alloys and Compounds,* 271(273), 78-84.

[18] Probian C., Wülfing, A., Harder, J. (2003). Anaerobic mineralization of quaternary carbon atoms: Isolation of denitrifying bacteria on pivalic acid (2,2-Dimethylpropionic acid), *Applied and Environmental Microbiology,* 69(3), 1866-1870.

[19] Vainshtein M., Suzina, N., Kudryashova, E., Ariskina, E. (2002). New Magnet-Sensitive Structures in Bacterial and Archaeal Cells, *Biol. Cell,* 94(1), 29-35.

[20] Tsagris E.M., de Alba, A.E., Gozmanova, M., Kalantidis, K. (2008). Viroids, *Cell Microbiol,* 10, 2168.

[21] Horie M., Honda, T., Suzuki, Y., Kobayashi, Y., Daito, T., Oshida, T. (2010). Endogenous non-retroviral RNA virus elements in mammalian genomes, *Nature,* 463, 84-87.

[22] Hecht M., Nitz, N., Araujo, P., Sousa, A., Rosa, A., Gomes, D. (2010). Genes from Chagas parasite can transfer to humans and be passed on to children. Inheritance of DNA Transferred from American Trypanosomes to Human Hosts, *PLoS ONE,* 5, 2-10.

[23] Flam F. (1994). Hints of a language in junk DNA, *Science,* 266, 1320.

[24] Horbach S., Sahm, H., Welle, R. (1993). Isoprenoid biosynthesis in bacteria: two different pathways? *FEMS Microbiol. Lett,* 111, 135–140.

[25] Gupta R.S. (1998). Protein phylogenetics and signature sequences: a reappraisal of evolutionary relationship among archaebacteria, eubacteria, and eukaryotes, *Microbiol. Mol. Biol. Rev,* 62, 1435–1491.

[26] Hanage W., Fraser, C., Spratt, B. (2005). Fuzzy species among recombinogenic bacteria, *BMC Biology,* 3, 6-10.

[27] Whitchurch C.B., Tolker-Nielsen, T., Ragas, P.C., Mattick, J.S. (2002). Extracellular DNA Required for Bacterial Biofilm Formation. *Science,* 295(5559), 1487.

[28] Webb J.S., Givskov, M., Kjelleberg, S. (2003). Bacterial biofilms: prokaryotic adventures in multicellularity, *Curr. Opin. Microbiol,* 6(6), 578–85.

[29] Chen Y., Cai, T., Wang, H., Li, Z., Loreaux, E., Lingrel, J.B. (2009). Regulation of intracellular cholesterol distribution by Na/K-ATPase, *J. Biol. Chem.,* 284(22), 14881-90.

[30] Poole A.M. (2006). Did group II intron proliferation in an endosymbiont-bearing archaeon create eukaryotes? *Biol. Direct,* 1, 36-40.

[31] Lockwood M. (1989). *Mind, Brain and the Quantum*. Oxford: B. Blackwell.

[32] Eberl M., Hintz, M., Reichenberg, A., Kollas, A., Wiesner, J., Jomaa, H. (2010). Microbial isoprenoid biosynthesis and human γδ T cell activation, *FEBS Letters,* 544(1), 4-10.

[33] Wallace D.C. (2005). Mitochondria and Cancer: Warburg Addressed, *Cold Spring Harbor Symposia on Quantitative Biology,* 70, 363-374.

[34] Lefebvre P., Cariou, B., Lien, F., Kuipers, F., Staels, B. (2009). Role of Bile Acids and Bile Acid Receptors in Metabolic Regulation, *Physiol Rev,* 89(1), 147 - 191.

[35] Russell M.J., Martin, W. (2004). The rocky roots of the acetyl-CoA Pathway, *Trends in Biochemical Sciences,* 29, 7.

[36] Margulis L. (1996). Archaeal-eubacterial mergers in the origin of Eukarya: phylogenetic classification of life, *Proc. Natl. Acad. Sci. USA,* 93, 1071-1076.

[37] Tielens A.G.G.M. (2008). Interstellar Polycyclic Aromatic Hydrocarbon Molecules, *Annual Review of Astronomy and Astrophysics,* 46, 289-337.

[38] Wickramasinghe C. (2004). The universe: a cryogenic habitat for microbial life, *Cryobiology,*48(2), 113-125.

[39] Hoyle F., Wickramasinghe, C. (1988). *Cosmic Life-Force.* London: J.M. Dent and Sons Ltd.

[40] Dun D. (2005). *The Black Silent.* New York: Pinnacle Books.

Chapter III

Archaeal Digoxin and Creation of Cellular Plasma State- Molecular/Cellular Electromagnetic Signal Transduction

Introduction

The human cell contains endosymbiotic archaea with ability to catabolize cholesterol and synthesise digoxin.[1] The nanoarchaea can get mineralized resulting in calcification and generation of cellular magnetite. The archaeal digoxin produces sodium potassium ATPase inhibition, a paroxysmal depolarisation shift in the neuronal membrane and generates an electromagnetic field of 10^{-7} Hz oscillation. Archaeal digoxin is thus involved in the generation of epileptic discharges. The electrical discharges induced by digoxin induced sodium potassium ATPase inhibition can generate a plasma state of matter within the cell. Thus archaeal digoxin can induce an electromagnetic field and a plasma state of matter intracellularly which may produce a electromagnetic and plasma correlate of molecules and cells.[2,3] The archaeal magnetite can also induce 10^{-7} oscillations. Electromagnetic oscillations in the range of 10^{-7} form the basis of consciousness. This can be the electromagnetic field or plasma state of matter related to human consciousness. The 10^{-7} electromagnetic oscillations are also seen in the intergalactic magnetic fields and Hoyle's postulate of a biological universe with intergalactic magnetotactic bacteria probably nanoarchaea is relevant in this context.[4,5,6] Archaeal digoxin by producing sodium potassium ATPase inhibition and a paroxysmal depolarisation shift can induce a pumped phonon system in dielectric molecules as existing in water, proteins and magnetite leading onto a superconducting frohlich state at normal temperature.[7] This forms the basis of digoxin induced quantal perception. The archaeal digoxin induced EMF field of 10^{-7} oscillations and quantal perceptive fields can generate electromagnetic traces of proteins, nucleic acids and carbohydrates. This EMF traces or co-resonance patterns of simple and complex molecules in the setting of cell oscillations can generate molecular memory in water. The piezoelectric effect induced by digoxin mediated sodium potassium ATPase inhibition can generate cellular succussion and the molecular electromagnetic traces can get stored in electric dipoles of water. These EMF traces of molecules or co-resonance pattern may form the basis of cell signal transduction. This has been demonstrated by the work of Montaigner with regard to DNA teleportation. The present experiment was designed to test the generation of electromagnetic signal traces of enzymes and their transmission in the setting of digoxin induced sodium potassium ATPase inhibition.

Materials and Methods

The setting of the experiment was designed as follows:

System I: Phosphate buffered saline (PBS)+ cholesterol substrate + plasma + rutile (0.1 mg/ml)+digoxin (0.5 ng/ml). System II: Control phosphate buffered saline (PBS). The system is incubated at 37°C for 1 hour and during the period of incubation both system I and II are kept close to each other. Then to control PBS add cholesterol substrate and incubate both the systems at 37°C for 16-18 hours. Cytochrome F420 was estimated flourimetrically (excitation wavelength 420 nm and emission wavelength 520 nm). The detection of cytochrome F420 indicates the presence of archaea. The fluorescence of cytochrome F420 was estimated at 0 time, after 1 hour and 16-18 hours in the system I and II. The test system I after 16-18 hours of incubation showed decrease in intensity of fluorescence due to destruction of archaea by long incubation consequent to exhaustion of substrate as well as digoxin induced cell destruction. There was a 21% increase in fluroscence in system II of PBS with added cholesterol substrate after 16-18 hours of incubation.

The system I of PBS + cholesterol substrate + plasma + rutile+digoxin had detectable cytochrome F420 indicating archaeal activity. In the presence of digoxin the system I produces membrane sodium potassium ATPase inhibition of archaeal membrane and a paroxysmal depolarisation shift. This generates an archaeal electromagnetic field of 10^{-7} Hz and a digoxin induced quantal pumped phonon system using archaeal dipolar magnetite. This generates an electromagnetic trace of the archaeal cell. The electromagnetic traces get dissolved in electric dipoles of water consequent to cellular movements and cytosol succussion induced by digoxin induced sodium potassium ATPase inhibition mediated piezoelectric effect. These archaeal cell electromagnetic signal traces get transmitted to system II of PBS + added cholesterol substrate. The electromagnetic traces of archaea generate cytochrome F420 in system II indicating the presence of archaea. The experiment indicates that in the presence of digoxin induced quantal pumped phonon system the electromagnetic trace of archaea can be generated and can be transmitted to the PBS+cholesterol substrate system resulting in generation of archaea/cytochrome F420 in the system. This could be a form of teleportation.

Discussion

The digoxin binding to sodium potassium ATPase of the archaeal membrane generates a paroxysmal depolarisation shift and an electromagnetic field of 10^{-7} Hz oscillations. The dipolar PAH and archaeal magnetite in the setting of digoxin induced sodium potassium

ATPase inhibition can produce a pumped phonon system mediated frohlich model superconducting state inducing quantal perception with nanoarchaeal sensed gravity producing the orchestrated reduction of the quantal possibilities to the macrosopic world. The archeal digoxin induced sodium potassium ATPase inhibition can produce a paroxysmal depolarisation shift and electric discharges. This can induce a plasma state of matter in cell. Plasma can exist as life forms with memory and capability to replicate. The initial life like forms during early evolution of universe plasma life forms with ability to self organise. The archaeal digoxin can generate an EMF signal, quantal signal and plasma signal of the cell, tissues and molecules. The human consciousness is postulated to exist as an electromagnetic field of 10^{-7} Hz oscillations or as plasma consciousness. Endosymbiotic archaeal digoxin can generate an electromagnetic or plasma co-resonance signal or trace of the cell, tissue and molecules. This electromagnetic or plasma signal is involved in signal transduction. This electromagnetic or plasma trace can have an extracorporeal existence and may be involved in extrasensory perception and quantal phenomena. The archaeal digoxin generated electromagnetic or plasma signal can get stored in electric dipoles of water consequent to cellular succussion generated by digoxin induced sodium potassium ATPase inhibition mediated piezoelectric effect. The digoxin induced macromolecular, cellular and tissue as well as corporeal electromagnetic and plasma signal can be transmitted to produce effects outside the cell, tissue and even body. The present experiment indicates the transfer of the cholesterol oxidase electromagnetic signal generated by digoxin to the system II to produce degradation of cholesterol generating H_2O_2. The results of the experiment correspond with the data on DNA teleportation reported by Montaigner.[2,3]

The 10^{-7} Hz oscillations are also noticed in the intergalactic magnetic field. The intergalactic magnetic field is generated by magnetotactic bacteria and is biological. This has been postulated by Fred Hoyle.[4,5,6] The intergalactic magnetic field is involved in the formation of star systems. The intergalactic colony of magnetotactic nanoarchaea generates the intergalactic magnetic field. The endosymbiotic nanoarchaea generates human consciousness by digoxin induced sodium potassium ATPase inhibition resulting in electromagnetic/plasma consciousness field.

The human brain and tissues can be visualized as being regulated by endosymbiotic archaeal colonies and having evolved from them. The archaeal digoxin forms the connecting link between the quantal/plasma world and the material body generating one from the other and connecting both together. The biological universe would have existed in the quantal/plasma state and the material universe would have evolved out of it due to the observer function of consciousness of plasma life forms. The plasma state of matter is self organizing and can exist as life forms with memory, intelligence and replicative ability.

The archaeal digoxin induced electromagnetic and plasma traces are eternal and can exists extracorporeally and forms the memory of nature as postulated by Sheldrake. The plasma state or electromagnetic consciousness field has past, present and future information stored in it as a form of quantum computer information storage system. The information contained in the consciousness field gets imprinted in the material body via digoxin induced quantal pumped phonon system.[7] The idea of molecular memory and electromagnetic/plasma molecular signal traces mediated signal transduction changes the concept of cell metabolism and function.

References

[1] Kurup R., Kurup, P.A. (2009). *Hypothalamic digoxin, cerebral dominance and brain function in health and diseases*. New York: Nova Science Publishers.

[2] Montagnier, L, Aäissa, J et al, "Electromagnetic Signals Are Produced by Aqueous Nanostructures Derived from Bacterial DNA Sequences". *Interdisciplinary Sciences: Computational Life Sciences*, 06 January 2009, Volume 1, Number 2, 81-90, DOI: 10.1007/s12539-009-0036-7 Accessed 27 Jan 2011.

[3] Montagnier, L, Aäissa, J et al, "Electromagnetic detection of HIV DNA in the blood of AIDS patients treated by antiretroviral therapy". *Interdisciplinary Sciences: Computational Life Sciences*, 03 July 2009, Volume 1, Number 4, 245-253, DOI: 10.1007/s12539-009-0059-0 Accessed 27 January 2011.

[4] From Grains to Bacteria, Hoyle, F. and Wickramasinghe N.C., University College Cardiff Press, ISBN 0906449642, 1984.

[5] Evolution from space (the Omni lecture) and other papers on the origin of life 1982, ISBN 0894900838.

[6] Evolution from Space: A Theory of Cosmic Creationism, 1984, ISBN 0-671-49263-2.

[7] Lockwood M. (1989). Mind, Brain and the Quantum. Oxford: B. Blackwell.

Chapter IV

An Actinide Dependent Shadow Biosphere of Archaea and Viroids in Neuropsychiatric Disorders

Introduction

Endomyocardial fibrosis (EMF) along with the root wilt disease of coconut is endemic to Kerala with its radioactive actinide beach sands. Actinides like rutile producing intracellular magnesium deficiency due to rutile-magnesium exchange sites in the cell membrane has been implicated in the etiology of EMF [1]. Endogenous digoxin, a steroidal glycoside which functions as a membrane sodium-potassium ATPase inhibitor has also been related to its etiology due to the intracellular magnesium deficiency it produces [2].

Organisms like phytoplasmas and viroids have also been demonstrated to play a role in the etiology of these diseases [3,4]. Endogenous digoxin has been related to the pathogenesis of schizophrenia, autism and primary seizure disorder [2]. The possibility of endogenous digoxin synthesis by actinide based primitive organism like archaea with a mevalonate pathway and cholesterol catabolism was considered [5,6,7].

Davies has put forward the concept of a shadow biosphere of organisms with alternate biochemistry present in earth itself [8]. An actinide dependent shadow biosphere of archaea and viroids in the above mentioned disease states is described [6].

Materials and Methods

Informed consent of the subjects and the approval of the ethics committee were obtained for the study. The following groups were included in the study:- schizophrenia, autism and primary seizure disorder/primary generalized epilepsy. There were 10 patients in each group and each patient had an age and sex matched healthy control selected randomly from the general population. The blood samples were drawn in the fasting state before treatment was initiated.

Table 1. Effect of rutile and antibiotics on muramic acid and dopamine

	DOPAMINE % (Increase with Rutile)		DOPAMINE % (Decrease with Doxy)		Muramic acid % change (Increase with Rutile)		Muramic acid % change (Decrease with Doxy+Cipro)	
Group	Mean	± SD	Mean	± SD	Mean	± SD	Mean	± SD
Normal	4.41	0.15	18.63	0.12	4.34	0.15	18.24	0.37
Schizo	21.88	1.19	66.28	3.60	23.02	1.65	67.61	2.77
Seizure	22.29	1.33	65.38	3.62	22.13	2.14	66.26	3.93
Autism	22.76	2.20	67.63	3.52	22.79	2.20	64.26	6.02
F value	403.394		680.284		348.867		364.999	
P value	< 0.001		< 0.001		< 0.001		< 0.001	

Table 2. Effect of rutile and antibiotics on free DNA and RNA

	DNA % change (Increase with Rutile)		DNA % change (Decrease with Doxy)		RNA % change (Increase with Rutile)		RNA % change (Decrease with Doxy)	
Group	Mean	± SD	Mean	± SD	Mean	± SD	Mean	± SD
Normal	4.37	0.15	18.39	0.38	4.37	0.13	18.38	0.48
Schizo	23.28	1.70	61.41	3.36	23.59	1.83	65.69	3.94
Seizure	23.40	1.51	63.68	4.66	23.08	1.87	65.09	3.48
Autism	22.12	2.44	63.69	5.14	23.33	1.35	66.83	3.27
F value	337.577		356.621		427.828		654.453	
P value	< 0.001		< 0.001		< 0.001		< 0.001	

Table 3. Effect of rutile and antibiotics on HMG CoA reductase and PAH

	HMG CoA R % change (Increase with Rutile)		HMG CoA R % change (Decrease with Doxy)		PAH % change (Increase with Rutile)		PAH % change (Decrease with Doxy)	
Group	Mean	± SD	Mean	± SD	Mean	± SD	Mean	± SD
Normal	4.30	0.20	18.35	0.35	4.45	0.14	18.25	0.72
Schizo	22.91	1.92	61.63	6.79	23.01	1.69	59.49	4.30
Seizure	23.09	1.69	61.62	8.69	22.67	2.29	57.69	5.29
Autism	22.72	1.89	64.51	5.73	22.61	1.42	64.48	6.90
F value	319.332		199.553		391.318		257.996	
P value	< 0.001		< 0.001		< 0.001		< 0.001	

Table 4. Effect of rutile and antibiotics on digoxin and bile acids

	Digoxin (ng/ml) (Increase with Rutile)		Digoxin (ng/ml) (Decrease with Doxy+Cipro)		Bile acids % change (Increase with Rutile)		Bile acids % change (Decrease with Doxy)	
Group	Mean	± SD	Mean	± SD	Mean	± SD	Mean	± SD
Normal	0.11	0.00	0.054	0.003	4.29	0.18	18.15	0.58
Schizo	0.55	0.06	0.219	0.043	23.20	1.87	57.04	4.27
Seizure	0.51	0.05	0.199	0.027	22.61	2.22	66.62	4.99
Autism	0.53	0.08	0.205	0.041	22.21	2.04	63.84	6.16
F value	135.116		71.706		290.441		203.651	
P value	< 0.001		< 0.001		< 0.001		< 0.001	

Table 5. Effect of rutile and antibiotics on pyruvate and hexokinase

	Pyruvate % change (Increase with Rutile)		Pyruvate % change (Decrease with Doxy)		Hexokinase % change (Increase with Rutile)		Hexokinase % change (Decrease with Doxy)	
Group	Mean	± SD	Mean	± SD	Mean	± SD	Mean	± SD
Normal	4.34	0.21	1[illegible].43	0.82	4.21	0.16	18.56	0.76
Schizo	20.99	1.46	6[illegible].23	9.73	23.01	2.61	65.87	5.27
Seizure	20.94	1.54	6[illegible].76	8.52	23.33	1.79	62.50	5.56
Autism	21.91	1.71	5[illegible].45	6.66	22.88	1.87	65.45	5.08
F value	321.255		1[illegible]5.242		292.065		317.966	
P value	< 0.001		< 0.001		< 0.001		< 0.001	

Table 6. Effect of rutile and antibiotics on hydrogen peroxide and delta amino levulinic acid

	H_2O_2 % (Increase with Rutile)		H_2O_2 % (Decrease with Doxy)		ALA % (Increase with Rutile)		ALA % (Decrease with Doxy)	
Group	Mean	± SD	Mean	± SD	Mean	± SD	Mean	± SD
Normal	4.43	0.19	13.13	0.63	4.40	0.10	18.48	0.39
Schizo	22.50	1.66	6[illegible].21	7.42	22.52	1.90	66.39	4.20
Seizure	23.81	1.19	6[illegible].08	7.38	22.83	1.90	67.23	3.45
Autism	23.52	1.49	6[illegible].24	7.36	23.20	1.57	66.65	4.26
F value	380.721		171.228		372.716		556.411	
P value	< 0.001		< 0.001		< 0.001		< 0.001	

Table 7. Effect of rutile and antibiotics on ATP synthase and cytochrome F 420

	ATP synthase % (Increase with Rutile)		ATP synthase % (Decrease with Doxy)		CYT F420 % (Increase with Rutile)		CYT F420 % (Decrease with Doxy)	
Group	Mean	± SD	Mean	± SD	Mean	± SD	Mean	± SD
Normal	4.40	0.11	18.78	0.11	4.48	0.15	18.24	0.66
Schizo	23.67	1.42	67.39	3.13	23.24	2.01	58.72	7.08
Seizure	23.09	1.90	66.15	4.09	23.46	1.87	59.27	8.86
Autism	22.60	1.64	66.86	4.21	21.68	1.90	57.93	9.64
F value	449.503		673.081		306.749		130.054	
P value	< 0.001		< 0.001		< 0.001		< 0.001	

Plasma from fasting heparinised blood was used and the experimental protocol was as follows (I) Plasma+phosphate buffered saline, (II) same as I+cholesterol substrate, (III) same as II+rutile 0.1 mg/ml, (IV) same as II+ciprofloxacine and doxycycline each in a concentration of 1 mg/ml. Cholesterol substrate was prepared as described by Richmond [9]. Aliquots were withdrawn at zero time immediately after mixing and after incubation at 37°C for 1 hour. The following estimations were carried out:- Cytochrome F420, free RNA, free DNA, muramic acid, polycyclic aromatic hydrocarbon, hydrogen peroxide, dopamine, pyruvate, ammonia, glutamate, cytochrome C, hexokinase, ATP synthase, HMG CoA redutase, digoxin and bile acids [10,11,12,13]. Cytochrome F420 was estimated flourimetrically (excitation wavelength 420 nm and emission wavelength 520 nm). Polycyclic aromatic hydrocarbon was estimated by measuring hydrogen peroxide liberated by using glucose reagent. The statistical analysis was done by ANOVA.

Results

The parameters checked as indicated above were:- cytochrome F420, free RNA, free DNA, muramic acid, polycyclic aromatic hydrocarbon, hydrogen peroxide, serotonin, pyruvate, ammonia, glutamate, cytochrome C, hexokinase, ATP synthase, HMG CoA reductase, digoxin and bile acids. Plasma of control subjects showed increased levels of the above mentioned parameters with after incubation for 1 hour and addition of cholesterol substrate resulted in still further significant increase in these parameters. The plasma of patients showed similar results but the extent of increase was more. The addition of antibiotics to the control plasma casued a decrease in all the parameters while addition of rutile increased their levels.

The addition of antibiotics to the patient's plasma caused a decrease in all the parameters while addition of rutile increased their levels but the extent of change was more in patient's sera as compared to controls. The results are expressed in tables 1-7 as percentage change in the parameters after 1 hour incubation as compared to the values at zero time.

Discussion

There was increase in cytochrome F420 indicating archaeal growth. The archaea can synthesise and use cholesterol as a carbon and energy source [14,15]. The archaeal origin of the enzyme activities was indicated by antibiotic induced suppression. The study indicates the presence of actinide based archaea with an alternate actinide based enzymes or metalloenzymes in the system as indicated by rutile induced increase in enzyme activities [16]. There was also an increase in archaeal HMG CoA reductase activity indicating increased cholesterol synthesis by the archaeal mevalonate pathway. The archaeal beta hydroxyl steroid dehydrogenase activity indicating digoxin synthesis and archaeal cholesterol hydroxylase activity indicating bile acid synthesis were increased [7]. The archaeal cholesterol oxidase activity was increased resulting in generation of pyruvate and hydrogen peroxide [15]. The pyruvate gets converted to glutamate and ammonia by the GABA shunt pathway. The archaeal aromatization of cholesterol generating PAH, serotonin and dopamine was also detected [17]. The archaeal glycolytic hexokinase activity and archaeal extracellular ATP synthase activity were increased. The archaea can undergo magnetite and calcium carbonate mineralization and can exist as calcified nanoforms [18]. There was an increase in free RNA indicating self replicating RNA viroids and free DNA indicating generation of viroid complementary DNA strands by archaeal reverse transcriptase activity. The actinides modulate RNA folding and catalyse its ribozymal action. Digoxin can cut and paste the viroidal strands by modulating RNA splicing generating RNA viroidal diversity. The viroids are evolutionarily escaped archaeal group I introns which have retrotransposition and self splicing qualities [19]. Archaeal pyruvate can produce histone deacetylase inhibition resulting in endogenous retroviral (HERV) reverse transcriptase and integrase expression. This can integrate the RNA viroidal complementary DNA into the noncoding region of eukaryotic non coding DNA using HERV integrase as has been described for borna and ebola viruses [20]. The noncoding DNA is lengthened by integrating RNA viroidal complementary DNA with the integration going on as a continuing event. The archaea genome can also get integrated into human genome using integrase as has been described for trypanosomes [21]. The integrated viroids and archaea can undergo vertical transmission and can exist as genomic parasites [20,21]. This increases the length and alters the grammar of the noncoding region producing memes or memory of acquired characters [22]. The viroidal complementary DNA can function as jumping genes producing a dynamic genome important in storage of synaptic information, HLA gene expression and neurodevelopmental gene expression. The alteration in DNA sequences produced by viroidal complementary DNA jumping genes can lead onto schizophrenia, autism and primary seizure disorder. The RNA viroids can regulate mrna function by RNA interference [19]. The phenomena of RNA interference can modulate T cell and B cell function, neuronal transmission and euchromatin/ heterochromatin expression. The RNA viroid induced mRNA interference can modulate dopaminergic, glutamatergic and serotoninergic synaptic transmission contributing to the pathogenesis of schizophrenia, autism and primary seizure disorder.

The presence of muramic acid, HMG CoA reductase and cholesterol oxidase activity inhibited by antibiotics indicates the presence of bacteria with mevalonate pathway. The bacterial with mevalonate pathway include streptococcus, staphylococcus, actinomycetes, listeria, coxiella and borrelia [23]. The bacteria and archaea with mevalonate pathway and cholesterol catabolism had a evolutionarily advantage and constitutes the isoprenoidal clade organism with the archaea evolving into mevalonate pathway gram positive and gram negative organism through horizontal gene transfer of viroidal and virus genes [24]. The isoprenoidal clade prokaryotes develop into other groups of prokaryotes via viroidal/virus as well as eukaryotic horizontal gene transfer producing bacterial speciation [25]. The RNA viroids and its complementary DNA developed into cholesterol enveloped RNA and DNA viruses like herpes, retrovirus, influenza virus, borna virus, cytomegalo virus and Ebstein Barr virus by recombining with eukaryotic and human genes resulting in viral speciation. Bacterial and viral species are ill defined and fuzzy with all of them forming one common genetic pool with frequent horizontal gene transfer and recombination. Thus the multi and unicellular eukaryote with its genes serves the purpose of prokaryotic and viral speciation .The multicellular eukaryote developed so that their endosymbiotic archaeal colonies could survive and forage better. The multicellular eukaryotes are like bacterial biofilms. The archaea and bacteria with a mevalonate pathway uses the extracellular RNA viroids and DNA viroids for quorum sensing and in the generation of symbiotic biofilm like structures which develop into multicellular eukaryotes [26,27]. The endosymbiotic archaea and bacteria with mevalonate pathway still uses the RNA viroids and DNA viroids for the regulation of muticellular eukaryote. Pollution is induced by the primitive nanoarchaea and mevalonate pathway bacteria synthesised PAH and methane leading on to redox stress. Redox stress leads to sodium potassium ATPase inhibition, inward movement of plasma membrane cholesterol, defective SREBP sensing, increased cholesterol synthesis and nanoarchaeal/mevalonate pathway bacterial growth [28]. Redox stress leads on to viroidal and archaeal multiplication. Redox stress can also lead to HERV reverse transcriptase and integrase expression. The noncoding DNA is formed of integrating RNA viroidal complementary DNA and archaea with the integration going on as a continuing event. The archaeal pox like dsDNA virus forms evolutionarily the nucleus. The integrated viroidal, archaeal and mevalonate pathway bacterial sequences can undergo vertical transmission and can exist as genomic parasites. The genomic integrated archaea, mevalonate pathway bacteria and viroids form a genomic reserve of bacteria and viruses which can recombine with human and eukaryotic genes producing bacterial and viral speciation. Bacteria and viruses have been related to the pathogenesis of schizophrenia, autism and primary seizure disorder. Borrelia, toxoplasma, chlamydia, mycoplasma, retroviruses, herpes virus, influenza virus and borna virus contribute to the neuropathogenesis of schizophrenia. Clostridia, rubella virus, influenza virus and borrelia are implicated in autism [29,30,31]. Herpes virus is implicated in the pathogenesis of primary seizure disorder. The change in the length and grammar of the noncoding region produces eukaryotic speciation and individuality [32]. Changes in the length of noncoding region can lead onto disorders of consciousness like schizophrenia [33]. The human endogenous retroviruses and change in the length and grammar of the noncoding region has been described in schizophrenia. The integration of nanoarchaea, mevalonate pathway prokaryotes and viroids in to the eukaryotic and human genome produces a chimera which can multiply producing biofilm like multicellular structures having a mixed archaeal, viroidal, prokaryotic and eukaryotic characters which is a regression from the multicellular eukaryotic tissue. This results in a new neuronal, metabolic, immune and tissue phenotype leading to human diseases like schizophrenia, autism and primary seizure disorder. The microchimeras formed can lead to polyploidy. Neuronal polyploidy and microchimeras have been described in schizophrenia and autism.

The archaea and viroids can regulate the nervous system including the NMDA/GABA thalamocorticothalamic pathway mediating conscious perception [2,34]. NMDA/GABA receptors can be modulated by digoxin induced calcium oscillations resulting in NMDA/glutamic acid decarboxylase (GAD) activity induction, PAH increasing NMDA activity and inducing GAD as well as viroid induced RNA interference modulating NMDA/GABA receptors [2]. The cholesterol ring oxidase generated pyruvate can be converted by the GABA shunt pathway to glutamate and GABA. Increased NMDA transmission has been described in schizophrenia, autism and primary seizure disorder. The dipolar PAH and archaeal magnetite in the setting of digoxin induced sodium potassium ATPase inhibition can produce a pumped phonon system mediated frohlich model superconducting state inducing quantal perception with nanoarchaeal sensed gravity producing the orchestrated reduction of the quantal possibilities to the macrosopic world [2,34]. Schizophrenia and autism are described as a disorder of consciousness and increased integration of archaea and viroids into the genome can contribute to its neuropathogenesis. The archaea can regulate limbic lobe transmission with archaeal cholesterol aromatase/ring oxidase generated norepinephrine, dopamine, serotonin and acetyl choline [17]. Increased dopaminergic and serotoninergic transmission is important in the pathogenesis of schizophrenia and autism. The higher degree of integration of the archaea into the genome produces increased digoxin synthesis producing right hemispheric dominance and lesser degree producing left hemispheric dominance [2]. Right hemispheric dominance has been described in schizophrenia, autism and primary seizure disorder. The increased integration of archaea into the neuronal genome can produce increased cholesterol oxidase and aromatase mediated monoamine and NMDA transmission producing schizophrenia, autism and primary seizure disorder. The archaeal bile acids are chemically diverse and structurally different from human bile acids. The archaeal bile acids can bind olfactory GPCR receptors and stimulate the limbic lobe producing a sense of social identity. The dominance of archaeal bile acids over human bile acids in stimulating the olfactory GPCR- limbic lobe pathway leads to loss of social identity and schizophrenia/ autism [35]. Archaea and RNA viroid can bind the TLR receptor induce NFKB producing immune activation and cytokine TNF alpha secretion. The archaeal DXP and mevalonate pathway metabolites can bind $\gamma\delta$ TCR and digoxin induced calcium signaling can activate NFKB producing chronic immune activation [2,36]. The archaea and viroid induced chronic immune activation and generation of superantigens can lead on to autoimmune disease. Immune activation and autoimmunity is important in the pathogenesis of schizophrenia, autism and primary seizure disorder. Autoantibodies have been described in schizophrenia, autism and primary seizure disorder. Archaea, viroids and digoxin can induce the host AKT PI3K, AMPK, HIF alpha and NFKB producing the Warburg metabolic phenotype [37]. The increased glycolytic hexokinase activity, decrease in blood ATP, leakage of cytochrome C, increase in serum pyruvate and decrease in acetyl CoA indicates the generation of the Warburg phenotype. There is induction of glycolysis, inhibition of PDH activity and mitochondrial dysfunction resulting in inefficient energetics. Mitochondrial dysfunction can lead onto NMDA excitotoxicity and cell death important in schizophrenia and primary seizure disorder. Cholesterol oxidase activity, increased glycolysis related NADPH oxidase activity and mitochondrial dysfunction generates free radicals important in the pathogenesis of schizophrenia, autism and primary seizure disorder. The accumulated pyruvate enters the GABA shunt pathway and is converted to citrate which is acted upon by citrate lyase and converted to acetyl CoA, used for cholesterol synthesis [37]. The pyruvate can be converted to glutamate and ammonia which is oxidised by archaea for energy needs. The increased cholesterol substrate leads to increased archaeal growth and digoxin synthesis leading to metabolic channeling to the mevalonate pathway. Hyperdigoxinemia is important in the pathogenesis of schizophrenia, autism and primary seizure disorder [2]. The Warburg phenotype can contribute to the pathogenesis of schizophrenia by augmenting the bacterial shikimic acid pathway. The upregulated glycolysis consequent to the Warburg phenotype produces phosphoenolpyruvate, a basic substrate for the bacterial shikimic acid pathway which can synthesise monoamines and neuroactive alkaloids. The shikimic acid pathway can generate dopamine and serotonin producing the increased monoaminergic transmission in schizophrenia. The shikimic acid pathway can also synthesise the neuroactive alkaloids strychnine, nicotine, morphine, mescaline and LSD important in the pathogenesis of schizophrenia and autism. Endogenous neuroactive alkaloids has been described in schizophrenia, autism and primary seizure disorder by several workers [2]. The upregulated glycolysis can also contribute to increased NMDA and GABA transmission in the thalamocorticothalamic pathway. The glycolytic pathway produces phosphoglycerate which is converted to phosphoserine and then serine which activates the NMDA receptor. The glycolytic enzyme glyceraldehyde-3-phosphate dehydrogenase is a GABA receptor kinase and activates GABA transmission. Thus the archaea and viroid induced Warburg phenotype can contribute to the pathogenesis of schizophrenia, autism and primary seizure disorder. The archaeal cholesterol catabolism can deplete the cell membranes of cholesterol resulting in alteration in lipid microdomains and their related neurotransmitter receptor contributing to the altered NMDA, serotoninergic and dopaminergic transmission in schizophrenia, autism and primary seizure disorder.

References

[1] Valiathan M.S., Somers, K., Kartha, C.C. (1993). *Endomyocardial Fibrosis.* Delhi: Oxford University Press.

[2] Kurup R., Kurup, P.A. (2009). *Hypothalamic digoxin, cerebral dominance and brain function in health and diseases.* New York: Nova Science Publishers.

[3] Hanold D., Randies, J.W. (1991). Coconut cadang-cadang disease and its viroid agent, *Plant Disease,* 75, 330-335.

[4] Edwin B.T., Mohankumaran, C. (2007). Kerala wilt disease phytoplasma: Phylogenetic analysis and identification of a vector, *Proutista moesta, Physiological and Molecular Plant Pathology,* 71(1-3), 41-47.

[5] Eckburg P.B., Lepp, P.W., Relman, D.A. (2003). Archaea and their potential role in human disease, *Infect Immun,* 71, 591-596.

[6] Adam Z. (2007). Actinides and Life's Origins, *Astrobiology,* 7, 6-10.

[7] Schoner W. (2002). Endogenous cardiac glycosides, a new class of steroid hormones, *Eur. J. Biochem.,* 269, 2440-2448.

[8] Davies P.C.W., Benner, S.A., Cleland, C.E., Lineweaver, C.H., McKay, C.P., Wolfe-Simon, F. (2009). Signatures of a Shadow Biosphere, *Astrobiology,* 10, 241-249.

[9] Richmond W. (1973). Preparation and properties of a cholesterol oxidase from nocardia species and its application to the enzymatic assay of total cholesterol in serum, *Clin. Chem.,* 19, 1350-1356.

[10] Snell E.D., Snell, C.T. (1961). *Colorimetric Methods of Analysis.* Vol 3A. New York: Van NoStrand.

[11] Glick D. (1971). *Methods of Biochemical Analysis.* Vol 5. New York: Interscience Publishers.

[12] Colowick, Kaplan, N.O. (1955). *Methods in Enzymology.* Vol 2. New York: Academic Press.

[13] Maarten A.H., Marie-Jose, M., Cornelia, G., van Helden–Meewsen, Fritz, E., Marten, P.H. (1995). Detection of muramic acid in human spleen, *Infection and Immunity,* 63(5), 1652 – 1657.

[14] Smit A., Mushegian, A. (2000). Biosynthesis of isoprenoids via mevalonate in Archaea: the lost pathway, *Genome Res,* 10(10), 1468-84.

[15] Van der Geize R., Yam, K., Heuser, T., Wilbrink, M.H., Hara, H., Anderton, M.C. (2007). A gene cluster encoding cholesterol catabolism in a soil actinomycete provides

insight into Mycobacterium tuberculosis survival in macrophages, *Proc. Natl. Acad. Sci. USA,* 104(6), 1947-52.

[16] Francis A.J. (1998). Biotransformation of uranium and other actinides in radioactive wastes, *Journal of Alloys and Compounds,* 271(273), 78-84.

[17] Probian C., Wülfing, A., Harder, J. (2003). Anaerobic mineralization of quaternary carbon atoms: Isolation of denitrifying bacteria on pivalic acid (2,2-Dimethylpropionic acid), *Applied and Environmental Microbiology,* 69(3), 1866-1870.

[18] Vainshtein M., Suzina, N., Kudryashova, E., Ariskina, E. (2002). New Magnet-Sensitive Structures in Bacterial and Archaeal Cells, *Biol. Cell,* 94(1), 29-35.

[19] Tsagris E.M., de Alba, A.E., Gozmanova, M., Kalantidis, K. (2008). Viroids, *Cell Microbiol.,* 10, 2168.

[20] Horie M., Honda, T., Suzuki, Y., Kobayashi, Y., Daito, T., Oshida, T. (2010). Endogenous non-retroviral RNA virus elements in mammalian genomes, *Nature,* 463, 84-87.

[21] Hecht M., Nitz, N., Araujo, P., Sousa, A., Rosa, A., Gomes, D. (2010). Genes from Chagas parasite can transfer to humans and be passed on to children. Inheritance of DNA Transferred from American Trypanosomes to Human Hosts, *PLoS ONE,* 5, 2-10.

[22] Flam F. (1994). Hints of a language in junk DNA, *Science,* 266, 1320.

[23] Horbach S., Sahm, H., Welle, R. (1993). Isoprenoid biosynthesis in bacteria: two different pathways? *FEMS Microbiol. Lett.,* 111, 135–140.

[24] Gupta R.S. (1998). Protein phylogenetics and signature sequences: a reappraisal of evolutionary relationship among archaebacteria, eubacteria, and eukaryotes, *Microbiol. Mol. Biol. Rev.,* 62, 1435–1491.

[25] Hanage W., Fraser, C., Spratt, B. (2005). Fuzzy species among recombinogenic bacteria, *BMC Biology,* 3, 6-10.

[26] Webb J.S., Givskov, M., Kjelleberg, S. (2003). Bacterial biofilms: prokaryotic adventures in multicellularity, *Curr. Opin. Microbiol,* 6(6), 578–85.

[27] Whitchurch C.B., Tolker-Nielsen, T., Ragas, P.C., Mattick, J.S. (2002). Extracellular DNA Required for Bacterial Biofilm Formation. *Science,* 295(5559), 1487.

[28] Chen Y., Cai, T., Wang, H., Li, Z., Loreaux, E., Lingrel, J.B. (2009). Regulation of intracellular cholesterol distribution by Na/K-ATPase, *J. Biol. Chem.,* 284(22), 14881-90.

[29] Fritzsche M. (2002). Seasonal correlation of sporadic schizophrenia to Ixodes ticks and Lyme borreliosis. *Int. J. Health Geogr,* 1(1), 2.

[30] Waltrip R.W. 2nd, Buchanan, R.W., Summerfelt, A., Breier, A., Carpenter, W.T. Jr., Bryant, N.L., Rubin, S.A., Carbone, K.M. (1995). Borna disease virus and schizophrenia. *Psych. Res.,* 56(1), 33-44.

[31] Torrey E.F., Yolken, R.H. (2003). Toxoplasma gondii and schizophrenia. *Emerg. Infect. Dis.,* 9(11), 1375-80.

[32] Poole A.M. (2006). Did group II intron proliferation in an endosymbiont-bearing archaeon create eukaryotes? *Biol. Direct,* 1, 36-40.

[33] Villarreal L.P. (2006). How viruses shape the tree of life, *Future Virology,* 1(5), 587-595.

[34] Lockwood M. (1989). *Mind, Brain and the Quantum*. Oxford: B. Blackwell.

[35] Lefebvre P., Cariou, B., Lien, F., Kuipers, F., Staels, B. (2009). Role of Bile Acids and Bile Acid Receptors in Metabolic Regulation, *Physiol. Rev,* 89(1), 147 - 191.

[36] Eberl M., Hintz, M., Reichenberg, A., Kollas, A., Wiesner, J., Jomaa, H. (2010). Microbial isoprenoid biosynthesis and human γδ T cell activation, *FEBS Letters,* 544(1), 4-10.

[37] Wallace D.C. (2005). Mitochondria and Cancer: Warburg Addressed, *Cold Spring Harbor Symposia on Quantitative Biology,* 70, 363-374.

Chapter V

A Cholesterol and Actinide Dependent Shadow Biosphere of Archaea and Viroids in Neurodegenerative Disorders

Abbreviations

HD: Huntington's disease
AD: Alzheimer's disease
MND: Motor neuron disease
PD: Parkinson's disease

Introduction

Endomyocardial fibrosis (EMF) along with the root wilt disease of coconut is endemic to Kerala with its radioactive actinide beach sands. Actinides like rutile producing intracellular magnesium deficiency due to rutile-magnesium exchange sites in the cell membrane has been implicated in the etiology of EMF [1]. Endogenous digoxin, a steroidal glycoside which functions as a membrane sodium-potassium ATPase inhibitor has also been related to its etiology due to the intracellular magnesium deficiency it produces.[2] Organisms like phytoplasmas and viroids have also been demonstrated to play a role in the etiology of these diseases.[3,4] Endogenous digoxin has been related to the pathogenesis of neuronal degenerations like parkinson's disease, alzheimer's disease, huntington's disease and motor neuron disease.[2]

The possibility of endogenous digoxin synthesis by actinide based primitive organism like archaea with a mevalonate pathway and cholesterol catabolism was considered.[5,6,7] Davies has put forward the concept of a shadow biosphere of organisms with alternate biochemistry present in earth itself. [8] An actinide dependent shadow biosphere of archaea and viroids in the above mentioned neuronal degenerations is described.

Materials and Methods

Informed consent of the subjects and the approval of the ethics committee were obtained for the study. The following groups were included in the study:- alzheimer's disease, parkinson's disease, huntington's disease and motor neuron disease. There were 10 patients in each group and each patient had an age and sex matched healthy control selected randomly from the general population.

The blood samples were drawn in the fasting state before treatment was initiated. Plasma from fasting heparinised blood was used and the experimental protocol was as follows (I) Plasma+phosphate buffered saline, (II) same as I+cholesterol substrate, (III) same as II+rutile 0.1 mg/ml, (IV) same as II+ciprofloxacine and doxycycline each in a concentration of 1 mg/ml. Cholesterol substrate was prepared as described by Richmond [9].

Aliquots were withdrawn at zero time immediately after mixing and after incubation at 37°C for 1 hour. The following estimations were carried out:- Cytochrome F420, free RNA, free DNA, polycyclic aromatic hydrocarbon, hydrogen peroxide, dopamine, pyruvate, ammonia, glutamate, cytochrome C, hexokinase, ATP synthase, HMG CoA redutase, digoxin and bile acids. [10,11,12]. Cytochrome F420 was estimated flourimetrically (excitation wavelength 420 nm and emission wavelength 520 nm). Polycyclic aromatic hydrocarbon was estimated by measuring hydrogen peroxide liberated by using glucose reagent. The statistical analysis was done by ANOVA.

Results

The parameters checked as indicated above were:- cytochrome F420, free RNA, free DNA, polycyclic aromatic hydrocarbon, hydrogen peroxide, serotonin, pyruvate, ammonia, glutamate, cytochrome C, hexokinase, ATP synthase, HMG CoA redutase, digoxin and bile acids. Plasma of control subjects showed increased levels of the above mentioned parameters with after incubation for 1 hour and addition of cholesterol substrate resulted in still further significant increase in these parameters. The plasma of patients showed similar results but the extent of increase was more. The addition of antibiotics to the control plasma caused a decrease in all the parameters while addition of rutile increased their levels. The addition of antibiotics to the patient's plasma caused a decrease in all the parameters while addition of rutile increased their levels but the extent of change was more in patient's sera as compared to controls. The results are expressed in tables 1-7 as percentage change in the parameters after 1 hour incubation as compared to the values at zero time.

Discussion

There was increase in cytochrome F420 indicating archaeal growth in motor neuron disease, parkinson's disease, huntington's disease and alzheimer's disease. The archaea can synthesise and use cholesterol as a carbon and energy source.[13,14] The archaeal origin of the enzyme activities was indicated by antibiotic induced suppression.

Table 1. Effect of rutile and antibiotics on muramic acid and dopamine

	DOPAMINE % (Increase with ce)		DOPAMINE % (Decrease with Doxy)		Muramic acid % (Increase without Doxy)		Muramic acid % (Decrease with Doxy)	
Normal	4.41	0.15	1[illegible].63	0.12	4.34	0.15	18.24	0.37
HD	23.79	1.58	6[illegible].56	4.03	22.30	2.19	66.19	4.20
AD	23.66	1.67	6[illegible].97	3.36	23.09	1.81	65.86	4.27
PD	23.21	1.74	6[illegible].76	3.15	22.48	2.13	63.12	4.84
MND	23.89	1.69	6[illegible].09	3.89	21.94	2.03	64.29	5.35
Aging	22.71	1.82	6[illegible].13	3.83	22.93	2.08	63.49	5.01
F value	403.394		[illegible]0.284		348.867		364.999	
P value	< 0.001		< 0.001		< 0.001		< 0.001	

Table 2. Effect of rutile and antibiotics on free DNA and RNA

	DNA % change (Increase with rutile)		DNA % change (Decrease with Doxy)		RNA % change (Increase with rutile)		RNA % change (Decrease with Doxy)	
Normal	4.37	0.15	[illegible].39	0.38	4.37	0.13	18.38	0.48
HD	22.48	2.13	[illegible]3.12	4.84	23.86	1.86	65.93	3.95
AD	23.52	1.65	[illegible]4.15	4.60	23.29	1.92	65.39	3.95
PD	22.30	2.19	[illegible]6.19	4.20	23.16	1.60	64.21	3.43
MND	23.11	2.00	[illegible]1.52	4.97	23.04	1.66	66.13	3.49
Aging	19.73	2.27	[illegible]5.49	7.28	19.73	2.27	62.70	3.24
F value	337.577		[illegible]56.621		427.828		654.453	
P value	< 0.001		< 0.001		< 0.001		< 0.001	

Table 3. Effect of rutile and antibiotics on HMG CoA reductase and PAH

	HMG CoA R % change (Increase with Rutile)		HMG CoA R % change (Decrease with Doxy)		PAH % change (Increase with Rutile)		PAH % change (Decrease with Doxy)	
Normal	4.30	0.20	18.35	0.35	4.45	0.14	18.25	0.72
HD	22.86	1.78	61.03	6.13	23.37	1.42	61.01	5.91
AD	23.43	1.68	61.68	8.32	23.26	1.53	60.91	7.59
PD	22.12	2.27	60.98	8.29	23.63	1.75	62.23	5.43
MND	21.79	1.68	64.51	6.96	23.17	2.02	61.03	5.40

Table 3. (Continued)

	HMG CoA R % change (Increase with Rutile)		HMG CoA R % change (Decrease with Doxy)		PAH % change (Increase with Rutile)		PAH % change (Decrease with Doxy)	
Aging	22.94	2.59	59.19	7.18	22.66	1.96	65.88	5.01
F value	319.332		199.553		391.318		257.996	
P value	< 0.001		< 0.001		< 0.001		< 0.001	

Table 4. Effect of rutile and antibiotics on digoxin and bile acids

	Digoxin (ng/ml) (Increase with Rutile)		Digoxin (ng/ml) (Decrease with Doxy+Cipro)		Bile acids % change (Increase with Rutile)		Bile acids % change (Decrease with Doxy)	
Normal	0.11	0.00	0.054	0.003	4.29	0.18	18.15	0.58
HD	0.52	0.09	0.177	0.038	23.08	1.56	62.00	5.39
AD	0.55	0.03	0.192	0.040	22.12	2.19	62.86	6.28
PD	0.54	0.03	0.193	0.042	23.77	1.40	65.39	4.88
MND	0.53	0.06	0.229	0.051	23.53	1.78	61.61	6.77
Aging	0.56	0.10	0.238	0.049	24.58	1.08	64.20	5.16
F value	135.116		71.706		290.441		203.651	
P value	< 0.001		< 0.001		< 0.001		< 0.001	

Table 5. Effect of rutile and antibiotics on pyruvate and hexokinase

	Pyruvate % change (Increase with Rutile)		Pyruvate % change (Decrease with Doxy)		Hexokinase % change (Increase with Rutile)		Hexokinase % change (Decrease with Doxy)	
Normal	4.34	0.21	18.43	0.82	4.21	0.16	18.56	0.76
HD	21.13	1.27	61.54	10.03	22.89	1.88	63.39	4.97
AD	22.63	0.88	56.40	8.59	22.96	2.12	65.11	5.91
PD	21.64	0.67	61.36	8.49	22.95	1.82	64.15	4.62
MND	21.58	0.81	59.11	10.05	23.15	1.78	64.41	4.90
Aging	21.31	2.51	60.42	7.65	23.36	1.78	66.62	4.83
F value	321.255		115.242		292.065		317.966	
P value	< 0.001		< 0.001		< 0.001		< 0.001	

Table 6. Effect of rutile and antibiotics on hydrogen peroxide and delta amino levulinic acid

	H_2O_2 % (Increase with ce)		H_2O_2 % (Decrease with Doxy)		ALA % (Increase with Rutile)		ALA % (Decrease with Doxy)	
Normal	4.43	0.19	13.13	0.63	4.40	0.10	18.48	0.39
HD	22.27	1.71	60.02	8.51	23.21	1.74	67.76	3.15
AD	22.65	2.48	60.19	6.98	23.67	1.68	66.50	3.58
PD	24.17	1.33	55.09	6.56	23.79	1.58	65.56	4.03
MND	23.58	1.94	57.85	6.63	23.06	1.72	64.82	3.31
Aging	22.27	1.87	61.77	6.79	19.73	2.27	64.78	6.62
F value	380.721		171.228		372.716		556.411	
P value	< 0.001		< 0.001		< 0.001		< 0.001	

Table 7. Effect of rutile and antibiotics on ATP synthase and cytochrome F 420

	ATP synthase % (Increase with ce)		ATP synthase % (Decrease with Doxy)		CYT F420 % (Increase with Rutile)		CYT F420 % (Decrease with Doxy)	
Normal	4.40	0.11	13.78	0.11	4.48	0.15	18.24	0.66
HD	23.16	1.60	64.21	3.43	22.10	2.83	59.72	6.90
AD	23.58	2.08	65.21	3.69	23.12	2.00	56.90	6.94
PD	23.86	1.86	65.93	3.95	22.32	2.17	57.31	9.22
MND	23.75	1.81	65.49	4.11	22.76	2.20	61.60	8.74
Aging	23.19	1.74	65.68	4.06	22.09	1.38	61.42	7.26
F value	449.503		673.081		306.749		130.054	
P value	< 0.001		< 0.001		< 0.001		< 0.001	

The study indicates the presence of actinide based archaea with an alternate actinide based enzymes or metalloenzymes in the system as indicated by rutile induced increase in enzyme activities.[15] There was also an increase in archaeal HMG CoA reductase activity indicating increased cholesterol synthesis by the archaeal mevalonate pathway. The archaeal beta hydroxyl steroid dehydrogenase activity indicating digoxin synthesis and archaeal cholesterol hydroxylase activity indicating bile acid synthesis were increased.[7] The archaeal cholesterol oxidase activity was increased resulting in generation of pyruvate and hydrogen peroxide.[14] The pyruvate gets converted to glutamate and ammonia by the GABA shunt pathway. The archaeal aromatization of cholesterol generating PAH, serotonin and dopamine was also detected.[16]

The archaeal glycolytic hexokinase activity and archaeal extracellular ATP synthase activity were increased. The archaea can undergo magnetite and calcium carbonate mineralization and can exist as calcified nanoforms.[17].

There was an increase in free RNA indicating self replicating RNA viroids and free DNA indicating generation of viroid complementary DNA strands by archaeal reverse transcriptase activity. The actinides modulate RNA folding and catalyse its ribozymal action. Digoxin can cut and paste the viroidal strands by modulating RNA splicing generating RNA viroidal diversity. The viroids are evolutionarily escaped archaeal group I introns which have retrotransposition and self splicing qualities.[18] Archaeal pyruvate can produce histone deacetylase inhibition resulting in endogenous retroviral (HERV) reverse transcriptase and integrase expression. This can integrate the RNA viroidal complementary DNA into the noncoding region of eukaryotic non coding DNA using HERV integrase as has been described for borna and ebola viruses.[19] The noncoding DNA is lengthened by integrating RNA viroidal complementary DNA with the integration going on as a continuing event. The archaea genome can also get integrated into human genome using integrase as has been described for trypanosomes.[20] The integrated viroids and archaea can undergo vertical transmission and can exist as genomic parasites.[19,20] This increases the length and alters the grammar of the noncoding region producing memes or memory of acquired characters.[21] The viroidal complementary DNA can function as jumping genes producing a dynamic genome important in storage of synaptic information, HLA gene expression and developmental gene expression. The RNA viroids can regulate mrna function by RNA interference.[18] The phenomena of RNA interference can modulate T cell and B cell function, apoptosis, neuronal transmission and euchromatin/heterochromatin expression. The phenomenon of RNA interference and the RNA viroidal complimentary DNA related jumping genes can lead onto proof reading errors and generation of trinucleotide repeats contributing to the pathogenesis of huntington's disease.

The presence of muramic acid, HMG CoA reductase and cholesterol oxidase activity inhibited by antibiotics indicates the presence of bacteria with mevalonate pathway. The bacterial with mevalonate pathway include streptococcus, staphylococcus, actinomycetes, listeria, coxiella and borrelia.[22]. The bacteria and archaea with mevalonate pathway and cholesterol catabolism had a evolutionarily advantage and constitutes the isoprenoidal clade organism with the archaea evolving into mevalonate pathway gram positive and gram negative organism through horizontal gene transfer of viroidal and virus genes [23] The isoprenoidal clade prokaryotes develop into other groups of prokaryotes via viroidal/virus as well as eukaryotic horizontal gene transfer producing bacterial speciation.[24] The RNA viroids and its complementary DNA developed into cholesterol enveloped RNA and DNA viruses like herpes, retrovirus, influenza virus, borna virus, cytomegalo virus and Ebstein Barr virus by recombining with eukaryotic and human genes resulting in viral speciation. Bacterial and viral species are ill defined and fuzzy with all of them forming one common genetic pool with frequent horizontal gene transfer and recombination. Thus the multi and unicellular eukaryote with its genes serves the purpose of prokaryotic and viral speciation .The multicellular eukaryote developed so that their endosymbiotic archaeal colonies could survive and forage better .The multicellular eukaryotes are like bacterial biofilms. The archaea and bacteria with a mevalonate pathway uses the extracellular RNA viroids and DNA viroids for quorum sensing and in the generation of symbiotic biofilm like structures which develop into multicellular eukaryotes [25,26]

The endosymbiotic archaea and bacteria with mevalonate pathway still uses the RNA viroids and DNA viroids for the regulation of muticellular eukaryote. Pollution is induced by the primitive nanoarchaea and mevalonate pathway bacteria synthesised PAH and methane leading on to redox stress. Redox stress leads to sodium potassium ATPase inhibition, inward movement of plasma membrane cholesterol, defective SREBP sensing, increased cholesterol synthesis and nanoarchaeal/mevalonate pathway bacterial growth.[27] Redox stress leads on to viroidal and archaeal multiplication. Redox stress can also lead to HERV reverse transcriptase and integrase expression. The noncoding DNA is formed of integrating RNA viroidal complementary DNA and archaea with the integration going on as a continuing event. The archaeal pox like dsDNA virus forms evolutionarily the nucleus .The integrated viroidal, archaeal and mevalonate pathway bacterial sequences can undergo vertical transmission and can exist as genomic parasites. The genomic integrated archaea, mevalonate pathway bacteria and viroids form a genomic reserve of bacteria and viruses which can recombine with human and eukaryotic genes producing bacterial and viral speciation. Bacteria and viruses have been related to the pathogenesis of motor neuron disease, alzheimer's disease and parkinson's disease. Chlamydia, mycoplasma, cyanobacteria, actinomycetes and borrelia have been reported to be involved in the pathogenesis of alzheimer's disease. [28,29,30] Helicobacter pylori, nocardia, streptococcus and corona viruses have been implicated in parkinson's disease [31,32] Mycoplasma, borrelia, retroviruses and enteroviruses have been related to the pathogenesis of MND [33,34] The change in the length and grammar of the noncoding region.[35] The integration of nanoarchaea, mevalonate pathway prokaryotes and viroids in to the eukaryotic and human genome produces a chimera which can multiply producing biofilm like multicellular structures having a mixed archaeal, viroidal, prokaryotic and eukaryotic characters which is a regression from the multicellular eukaryotic tissue. This results in a new neuronal, metabolic, immune and tissue phenotype leading to human diseases like neuronal degeneration. The microchimeras formed can lead to polyploidy which has been implicated in degenerations like alzheimer's disease. Microchimeras can lead onto autoimmune disease.

The archaea and viroids can regulate the nervous system including the NMDA transmission.[2] NMDA receptors can be activated by digoxin induced calcium oscillations, PAH increasing NMDA activity as well as viroid induced RNA interference [2] The cholesterol ring oxidase generated pyruvate can be converted by the GABA shunt pathway to glutamate contributing to NMDA excitotoxicity. The archaea can regulate dopaminergic transmission with archaeal cholesterol aromatase/ring oxidase generated dopamine [16] The increased dopamine synthesis can generate increased free radicals consequent to its catabolism. Cholesterol oxidase can generate free radical hydrogen peroxide. Free radicals can produce neuronal degeneration. The higher degree of integration of the archaea into the genome produces increased digoxin synthesis producing right hemispheric dominance and lesser degree producing left hemispheric dominance [2] Previous studies by the authors have related right hemispheric chemical dominance to neuronal degeneration. Archaea and RNA viroid can bind the TLR receptor induce NFKB producing immune activation and cytokine TNF alpha secretion. The archaeal DXP and mevalonate pathway metabolites can bind $\gamma\delta$ TCR and digoxin induced calcium signaling can activate NFKB producing chronic immune activation [2,36] The archaea and viroid induced chronic immune activation and generation of superantigens can lead on to autoimmune disease. Immune activation and autoantibodies have been related to neuronal degeneration. Immune activation and free radicals induce neutral sphingomyelinase generating ceramide. Ceramide acts upon the mitochondrial PT pore producing cell death. Archaea, viroids and digoxin can induce the host AKT PI3K, AMPK, HIF alpha and NFKB producing the Warburg metabolic phenotype [37] The increased glycolytic hexokinase activity, decrease in blood ATP, leakage of cytochrome C, increase in serum pyruvate and decrease in acetyl CoA indicates the generation of the Warburg phenotype. There is induction of glycolysis, inhibition of PDH activity and mitochondrial dysfunction resulting in inefficient energetics. Mitochondrial dysfunction has been related to neuronal degeneration. The increased glycolysis results in increased generation of the enzyme glyceraldehyde 3 phosphate dehydrogenase (GAPD). GAPD can undergo polyadenylation via free radical activated PARP enzyme. The polyadenylated GAPD can undergo nuclear translocation producing nuclear cell death. The accumulated pyruvate enters the gaba shunt pathway and is converted to citrate which is acted upon by citrate lyase and converted to acetyl CoA, used for cholesterol synthesis [37] The pyruvate can be converted to glutamate and ammonia which is oxidised by archaea for energy needs. Ammonia can produce NMDA excitotoxicity and cell death. Ammonia can activate sodium potassium ATPase producing increased neuronal requirement of ATP leading onto mitochondrial transmembrane potential changes and cell death. The increased cholesterol substrate leads to increased archaeal growth and digoxin synthesis leading to metabolic channelling to the mevalonate pathway. Digoxin can produce sodium potassium ATPase inhibition and increase in intracellular calcium producing mitochondrial PT pore dysfunction and cell death [2]. The archaeal cholesterol catabolism generated PAH can produce NMDA excitoxicity and cell death. The archaeal and mevalonate pathway bacteria cholesterol catabolism can deprive cholesterol from neuronal cell membrane and organelle membranes like mitochondrial, ER and lysosomal membranes producing cellular and organelle dysfunction and death. Cholesterol metabolic defect has been described in huntington's disease. Thus, the shadow biosphere of actinide dependent archaea, viroids and mevalonate pathway bacteria can lead onto neuronal degenerations like alzheimer's disease, huntington's disease, parkinson's disease and motor neuron disease.

References

[1] Valiathan M.S., Somers, K., Kartha, C.C. (1993). *Endomyocardial Fibrosis*. Delhi: Oxford University Press.

[2] Kurup R., Kurup, P.A. (2009). *Hypothalamic digoxin, cerebral dominance and brain function in health and diseases*. New York: Nova Science Publishers.

[3] Hanold D., Randies, J.W. (1991). Coconut cadang-cadang disease and its viroid agent, *Plant Disease,* 75, 330-335.

[4] Edwin B.T., Mohankumaran, C. (2007). Kerala wilt disease phytoplasma: Phylogenetic analysis and identification of a vector, *Proutista moesta, Physiological and Molecular Plant Pathology,* 71(1-3), 41-47.

[5] Eckburg P.B., Lepp, P.W., Relman, D.A. (2003). Archaea and their potential role in human disease, *Infect Immun,* 71, 591-596.

[6] Adam Z. (2007). Actinides and Life's Origins, *Astrobiology,* 7, 6-10.

[7] Schoner W. (2002). Endogenous cardiac glycosides, a new class of steroid hormones, *Eur. J. Biochem.,* 269, 2440-2448.

[8] Davies P.C.W., Benner, S.A., Cleland, C.E., Lineweaver, C.H., McKay, C.P., Wolfe-Simon, F. (2009). Signatures of a Shadow Biosphere, *Astrobiology,* 10, 241-249.

[9] Richmond W. (1973). Preparation and properties of a cholesterol oxidase from nocardia species and its application to the enzymatic assay of total cholesterol in serum, *Clin. Chem.,* 19, 1350-1356.

[10] Snell E.D., Snell, C.T. (1961). *Colorimetric Methods of Analysis.* Vol 3A. New York: Van NoStrand.

[11] Glick D. (1971). *Methods of Biochemical Analysis.* Vol 5. New York: Interscience Publishers.

[12] Colowick, Kaplan, N.O. (1955). *Methods in Enzymology.* Vol 2. New York: Academic Press.

[13] Smit A., Mushegian, A. (2000). Biosynthesis of isoprenoids via mevalonate in Archaea: the lost pathway, *Genome Res.,* 10(10), 1468-84.

[14] Van der Geize R., Yam, K., Heuser, T., Wilbrink, M.H., Hara, H., Anderton, M.C. (2007). A gene cluster encoding cholesterol catabolism in a soil actinomycete provides insight into Mycobacterium tuberculosis survival in macrophages, *Proc. Natl. Acad. Sci. USA,* 104(6), 1947-52.

[15] Francis A.J. (1998). Biotransformation of uranium and other actinides in radioactive wastes, *Journal of Alloys and Compounds,* 271(273), 78-84.

[16] Probian C., Wülfing, A., Harder, J. (2003). Anaerobic mineralization of quaternary carbon atoms: Isolation of denitrifying bacteria on pivalic acid (2,2-Dimethylpropionic acid), *Applied and Environmental Microbiology,* 69(3), 1866-1870.

[17] Vainshtein M., Suzina, N., Kudryashova, E., Ariskina, E. (2002). New Magnet-Sensitive Structures in Bacterial and Archaeal Cells, *Biol Cell,* 94(1), 29-35.

[18] Tsagris E.M., de Alba, A.E., Gozmanova, M., Kalantidis, K. (2008). Viroids, *Cell Microbiol.,* 10, 2168.

[19] Horie M., Honda, T., Suzuki, Y., Kobayashi, Y., Daito, T., Oshida, T. (2010). Endogenous non-retroviral RNA virus elements in mammalian genomes, *Nature,* 463, 84-87.

[20] Hecht M., Nitz, N., Araujo, P., Sousa, A., Rosa, A., Gomes, D. (2010). Genes from Chagas parasite can transfer to humans and be passed on to children. Inheritance of DNA Transferred from American Trypanosomes to Human Hosts, *PLoS ONE,* 5, 2-10.

[21] Flam F. (1994). Hints of a language in junk DNA, *Science,* 266, 1320.

[22] Horbach S., Sahm, H., Welle, R. (1993). Isoprenoid biosynthesis in bacteria: two different pathways? *FEMS Microbiol. Lett,* 111, 135–140.

[23] Gupta R.S. (1998). Protein phylogenetics and signature sequences: a reappraisal of evolutionary relationship among archaebacteria, eubacteria, and eukaryotes, *Microbiol. Mol. Biol. Rev.,* 62, 1435–1491.

[24] Hanage W., Fraser, C., Spratt, B. (2005). Fuzzy species among recombinogenic bacteria, *BMC Biology,* 3, 6-10.

[25] Whitchurch C.B., Tolker-Nielsen, T., Ragas, P.C., Mattick, J.S. (2002). Extracellular DNA Required for Bacterial Biofilm Formation. *Science,* 295(5559), 1487.

[26] Webb J.S., Givskov, M., Kjelleberg, S. (2003). Bacterial biofilms: prokaryotic adventures in multicellularity, *Curr. Opin. Microbiol,* 6(6), 578–85.

[27] Chen Y., Cai, T., Wang, H., Li, Z., Loreaux, E., Lingrel, J.B. (2009). Regulation of intracellular cholesterol distribution by Na/K-ATPase, *J. Biol. Chem.,* 284(22), 14881-90.

[28] MacDonald A.B. (2006). Concurrent Neocortical Borreliosis and Alzheimer's Disease. Demonstration of a Spirochetal Cyst Form, *Annals of the New York Academy of Sciences,* 539, 468 – 470.

[29] Howard J., Pilkington, G.J. (1992). Fibronectin staining detects micro-organisms in aged and Alzheimer's disease brain. *Neuroreport,* 3(7), 615-8.

[30] Balin B.J., Hammond, C.J., ScottLittle, C., MacIntyre, A., Appelt, D.M. (2000). *Chlamydia pneumoniae Infection and Disease.* New York: Kluwer Academic Publishers.

[31] Chapman G. (2003). In Situ Hybridization for Detection of Nocardial 16S rRNA: Reactivity within Intracellular Inclusions in Experimentally Infected Cynomolgus Monkeys-and in Lewy Body-containing Human Brain Specimens, *Experimental Neurology,* 184, 715-725.

[32] Dobbs R.J., Dobbs, S.M., Bjarnason, I.T. (2005) Role of chronic infection and inflammation in the gastroinstestinal tract in the etiology and pathogenesis of idiopathic parkinsonism. Part 1: Eradication of Helicobacter in the cachexia of idiopathic Parkinsonism, *Helicobact,* 10, 267–275.

[33] Halperin J.J., Kaplan, G.P., Brazinsky, S. (1990). Immunologic reactivity against Borrelia burgdorferi in patients with motor neuron disease. *Arch. Neurol.,* 47, 586–594.

[34] Ince P.G., Codd, G.A. (2005). Return of the cycad hypothesis–Does the amyotrophic lateral sclerosis/ Parkinsonism dementia complexes (ALS/PDC) of Guam has new implications for global health? *Neuropathol. Appl. Neurobiol.,* 31, 345–353.

[35] Poole A.M. (2006). Did group II intron proliferation in an endosymbiont-bearing archaeon create eukaryotes? *Biol. Direct,* 1, 36-40.

[36] Eberl M., Hintz, M., Reichenberg, A., Kollas, A., Wiesner, J., Jomaa, H. (2010). Microbial isoprenoid biosynthesis and human γδ T cell activation, *FEBS Letters,* 544(1), 4-10.

[37] Wallace D.C. (2005). Mitochondria and Cancer: Warburg Addressed, *Cold Spring Harbor Symposia on Quantitative Biology,* 70, 363-374.

Chapter VI

A Cholesterol and Actinide Dependent Shadow Biosphere of Archaea and Viroids in Neurodevelopmental Disorders

Introduction

Endomyocardial fibrosis (EMF) along with the root wilt disease of coconut is endemic to Kerala with its radioactive actinide beach sands. Actinides like rutile producing intracellular magnesium deficiency due to rutile-magnesium exchange sites in the cell membrane has been implicated in the etiology of EMF [1]. Endogenous digoxin, a steroidal glycoside which functions as a membrane sodium-potassium ATPase inhibitor has also been related to its etiology due to the intracellular magnesium deficiency it produces[2]. Organisms like phytoplasmas and viroids have also been demonstrated to play a role in the etiology of these diseases [3,4]. Endogenous digoxin has been related to the pathogenesis of autism, cerebral palsy and trisomy 21 [2]. The possibility of endogenous digoxin synthesis by actinide based primitive organism like archaea with a mevalonate pathway and cholesterol catabolism was considered [5,6,7]. Davies has put forward the concept of a shadow biosphere of organisms with alternate biochemistry present in earth itself [8]. An actinide dependent shadow biosphere of archaea and viroids in the above mentioned disease states is described [6].

Materials and Methods

Informed consent of the subjects and the approval of the ethics committee were obtained for the study. The following groups were included in the study:- Autism, cerebral palsy and trisomy 21. There were 10 patients in each group and each patient had an age and sex matched healthy control selected randomly from the general population. The blood samples were drawn in the fasting state before treatment was initiated. Plasma from fasting heparinised blood was used and the experimental protocol was as follows (I)

Plasma+phosphate buffered saline, (II) same as I+cholesterol substrate, (III) same as II+rutile 0.1 mg/ml, (IV) same as II+ciprofloxacine and doxycycline each in a concentration of 1 mg/ml. Cholesterol substrate was prepared as described by Richmond [9].

Aliquots were withdrawn at zero time immediately after mixing and after incubation at 37°C for 1 hour. The following estimations were carried out:- Cytochrome F420, free RNA, free DNA, muramic acid, polycyclic aromatic hydrocarbon, hydrogen peroxide, serotonin, pyruvate, ammonia, glutamate, cytochrome C, hexokinase, ATP synthase, HMG CoA redutase, digoxin and bile acids [10,11,12,13]. Cytochrome F420 was estimated flourimetrically (excitation wavelength 420 nm and emission wavelength 520 nm).

Polycyclic aromatic hydrocarbon was estimated by measuring hydrogen peroxide liberated by using glucose reagent. The statistical analysis was done by ANOVA.

Results

The parameters checked as indicated above were:- cytochrome F420, free RNA, free DNA, muramic acid, polycyclic aromatic hydrocarbon, hydrogen peroxide, serotonin, pyruvate, ammonia, glutamate, cytochrome C, hexokinase, ATP synthase, HMG CoA reductase, digoxin and bile acids. Plasma of control subjects showed increased levels of the above mentioned parameters with after incubation for 1 hour and addition of cholesterol substrate resulted in still further significant increase in these parameters. The plasma of patients showed similar results but the extent of increase was more. The addition of antibiotics to the control plasma caused a decrease in all the parameters while addition of rutile increased their levels.

The addition of antibiotics to the patient's plasma caused a decrease in all the parameters while addition of rutile increased their levels but the extent of change was more in patient's sera as compared to controls. The results are expressed in tables 1-7 as percentage change in the parameters after 1 hour incubation as compared to the values at zero time.

Discussion

There was increase in cytochrome F420 indicating archaeal growth. The archaea can synthesise and use cholesterol as a carbon and energy source [14,15]. The archeal origin of the enzyme activities was indicated by antibiotic induced suppression. The study indicates the presence of actinide based archaea with an alternate actinide based enzymes or metalloenzymes in the system as indicated by rutile induced increase in enzyme activities [16]. There was also an increase in archaeal HMG CoA reductase activity indicating increased cholesterol synthesis by the archaeal mevalonate pathway.

The archaeal beta hydroxyl steroid dehydrogenase activity indicating digoxin synthesis and archaeal cholesterol hydroxylase activity indicating bile acid synthesis were increased [7]. The archaeal cholesterol oxidase activity was increased resulting in generation of pyruvate and hydrogen peroxide [15].

Table 1. Effect of rutile and antibiotics on muramic acid and serotonin

	Muramic acid % change (Increase with Rutile)		Muramic acid % change (Decrease with Doxy+Cipro)		5 HT % (Increase without Doxy)		5 HT % (Decrease with Doxy)	
Normal	4.41	0.15	18.63	0.12	4.34	0.15	18.24	0.37
Autism	22.76	2.20	67.63	3.52	22.79	2.20	64.26	6.02
Trisomy 21	24.00	1.64	66.04	4.36	22.52	2.06	66.09	5.73
Cerebral Palsy	23.98	1.72	66.76	4.01	22.78	1.94	63.06	6.20
F value	403.394		680.284		348.867		364.999	
P value	< 0.001		< 0.001		< 0.001		< 0.001	

Table 2. Effect of rutile and antibiotics on free DNA and RNA

	DNA % change (Increase with Rutile)		DNA % change (Decrease with Doxy)		RNA % change (Increase with Rutile)		RNA % change (Decrease with Doxy)	
Normal	4.37	0.15	18.39	0.38	4.37	0.13	18.38	0.48
Autism	22.12	2.44	63.69	5.14	23.33	1.35	66.83	3.27
Trisomy 21	23.49	1.19	64.63	6.58	23.22	1.35	66.42	4.21
Cerebral Palsy	23.13	1.78	64.88	4.96	23.27	1.56	66.43	3.77
F value	337.577		356.621		427.828		654.453	
P value	< 0.001		< 0.001		< 0.001		< 0.001	

Table 3. Effect of rutile and antibiotics on HMG CoA reductase and PAH

	HMG CoA R % change (Increase with Rutile)		HMG CoA R % change (Decrease with Doxy)		PAH % change (Increase with Rutile)		PAH % change (Decrease with Doxy)	
Normal	4.30	0.20	18.35	0.35	4.45	0.14	18.25	0.72
Autism	22.72	1.89	64.51	5.73	22.61	1.42	64.48	6.90
Trisomy 21	23.82	1.78	61.63	7.96	24.06	1.50	63.01	7.76
Cerebral Palsy	22.44	2.27	63.92	4.99	23.06	1.78	65.73	4.00
F value	319.332		199.553		391.318		257.996	
P value	< 0.001		< 0.001		< 0.001		< 0.001	

Table 4. Effect of rutile and antibiotics on digoxin and bile acids

	Digoxin (ng/ml) (Increase with Rutile)		Digoxin (ng/ml) (Decrease with Doxy+Cipro)		Bile acids % change (Increase with Rutile)		Bile acids % change (Decrease with Doxy)	
Normal	0.11	0.00	0.054	0.003	4.29	0.18	18.15	0.58
Autism	0.53	0.08	0.205	0.041	22.21	2.04	63.84	6.16
Trisomy 21	0.52	0.08	0.208	0.031	21.65	2.37	65.91	4.82
Cerebral Palsy	0.51	0.07	0.212	0.034	22.18	2.40	65.89	6.18
F value	135.116		71.706		290.441		203.651	
P value	< 0.001		< 0.001		< 0.001		< 0.001	

Table 5. Effect of rutile and antibiotics on pyruvate and hexokinase

	Pyruvate % change (Increase with Rutile)		Pyruvate % change (Decrease with Doxy)		Hexokinase % change (Increase with Rutile)		Hexokinase % change (Decrease with Doxy)	
Normal	4.34	0.21	18.43	0.82	4.21	0.16	18.56	0.76
Autism	21.91	1.71	58.45	6.66	22.88	1.87	65.45	5.08
Trisomy 21	20.95	1.35	60.24	7.16	22.88	1.98	65.45	5.49
Cerebral Palsy	21.81	1.03	63.83	5.41	22.52	1.93	65.61	6.20
F value	321.255		115.242		292.065		317.966	
P value	< 0.001		< 0.001		< 0.001		< 0.001	

Table 6. Effect of rutile and antibiotics on hydrogen peroxide and delta amino levulinic acid

	H_2O_2 % (Increase with Rutile)		H_2O_2 % (Decrease with Doxy)		ALA % (Increase with Rutile)		ALA % (Decrease with Doxy)	
Normal	4.43	0.19	18.13	0.63	4.40	0.10	18.48	0.39
Autism	23.52	1.49	63.24	7.36	23.20	1.57	66.65	4.26
Trisomy 21	23.76	1.48	62.14	6.76	23.41	1.55	66.36	4.31
Cerebral Palsy	23.87	1.41	61.14	6.93	23.11	1.82	66.96	3.79
F value	380.721		171.228		372.716		556.411	
P value	< 0.001		< 0.001		< 0.001		< 0.001	

Table 7. Effect of rutile and antibiotics on ATP synthase and cytochrome F 420

	ATP synthase % (Increase with Rutile)		ATP synthase % (Decrease with Doxy)		CYT F420 % (Increase with Rutile)		CYT F420 % (Decrease with Doxy)	
Normal	4.40	0.11	18.78	0.11	4.48	0.15	18.24	0.66
Autism	22.60	1.64	66.86	4.21	21.68	1.90	57.93	9.64
Trisomy 21	23.34	1.58	65.76	3.91	22.78	2.19	58.97	8.84
Cerebral Palsy	22.91	1.69	66.23	3.44	22.84	2.66	55.33	6.05
F value P value	449.503 < 0.001		673.081 < 0.001		306.749 < 0.001		130.054 < 0.001	

The pyruvate gets converted to glutamate and ammonia by the GABA shunt pathway. The archaeal aromatization of cholesterol generating PAH, serotonin and dopamine was also detected [17]. The archaeal glycolytic hexokinase activity and archaeal extracellular ATP synthase activity were increased. The archaea can undergo magnetite and calcium carbonate mineralization and can exist as calcified nanoforms [18]. There was an increase in free RNA indicating self replicating RNA viroids and free DNA indicating generation of viroid complementary DNA strands by archaeal reverse transcriptase activity. The actinides modulate RNA folding and catalyse its ribozymal action. Digoxin can cut and paste the viroidal strands by modulating RNA splicing generating RNA viroidal diversity. The viroids are evolutionarily escaped archaeal group I introns which have retrotransposition and self splicing qualities [19]. Archaeal pyruvate can produce histone deacetylase inhibition resulting in endogenous retroviral (HERV) reverse transcriptase and integrase expression. This can integrate the RNA viroidal complementary DNA into the noncoding region of eukaryotic non coding DNA using HERV integrase as has been described for borna and ebola viruses [20]. The noncoding DNA is lengthened by integrating RNA viroidal complementary DNA with the integration going on as a continuing event. The archaea genome can also get integrated into human genome using integrase as has been described for trypanosomes [21]. The integrated viroids and archaea can undergo vertical transmission and can exist as genomic parasites [20,21]. This increases the length and alters the grammar of the noncoding region producing memes or memory of acquired characters [22]. The viroidal complementary DNA can function as jumping genes producing a dynamic genome important in storage of synaptic information, HLA gene expression and developmental gene expression. The RNA viroids can regulate mRNA function by RNA interference [19]. The phenomena of RNA interference can modulate T cell and B cell function, insulin signaling lipid metabolism, cell growth and differentiation, apoptosis, neuronal transmission and euchromatin/ heterochromatin expression.

The presence of muramic acid, HMG CoA reductase and cholesterol oxidase activity inhibited by antibiotics indicates the presence of bacteria with mevalonate pathway. The bacterial with mevalonate pathway include streptococcus, staphylococcus, actinomycetes, listeria, coxiella and borrelia[23]. The bacteria and archaea with mevalonate pathway and cholesterol catabolism had a evolutionarily advantage and constitutes the isoprenoidal clade organism with the archaea evolving into mevalonate pathway gram positive and gram negative organism through horizontal gene transfer of viroidal and virus genes [24]. The isoprenoidal clade prokaryotes develop into other groups of prokaryotes via viroidal/ virus as well as eukaryotic horizontal gene transfer producing bacterial speciation [25]. The RNA viroids and its complementary DNA developed into cholesterol enveloped RNA and DNA viruses like herpes, retrovirus, influenza virus, borna virus, cytomegalo virus and Ebstein Barr virus by recombining with eukaryotic and human genes resulting in viral speciation. Bacterial and viral species are ill defined and fuzzy with all of them forming one common genetic pool with frequent horizontal gene transfer and recombination. Thus the multi and unicellular eukaryote with its genes serves the purpose of prokaryotic and viral speciation. The multicellular eukaryote developed so that their endosymbiotic archaeal colonies could survive and forage better. The multicellular eukaryotes are like bacterial biofilms. The archaea and bacteria with a mevalonate pathway uses the extracellular RNA viroids and DNA viroids for quorum sensing and in the generation of symbiotic biofilm like structures which develop into multicellular eukaryotes [26,27]. The endosymbiotic archaea and bacteria with mevalonate pathway still uses the RNA viroids and DNA viroids for the regulation of muticellular eukaryote. Pollution is induced by the primitive nanoarchaea and mevalonate pathway bacteria synthesised PAH and methane leading on to redox stress. Redox stress leads to sodium potassium ATPase inhibition, inward movement of plasma membrane cholesterol, defective SREBP sensing, increased cholesterol synthesis and nanoarchaeal/mevalonate pathway bacterial growth [28]. Redox stress leads on to viroidal and archaeal multiplication. Redox stress can also lead to HERV reverse transcriptase and integrase expression. The noncoding DNA is formed of integrating RNA viroidal complementary DNA and archaea with the integration going on as a continuing event. The archaeal pox like dsDNA virus forms evolutionarily the nucleus .The integrated viroidal, archaeal and mevalonate pathway bacterial sequences can undergo vertical transmission and can exist as genomic parasites. The genomic integrated archaea, mevalonate pathway bacteria and viroids form a genomic reserve of bacteria and viruses which can recombine with human and eukaryotic genes producing bacterial and viral speciation. Bacteria and viruses have been related to the pathogenesis of autism. Clostridia, mycoplasma and borrelia have been related to autism [29]. Measles virus, influenza virus and borna virus are important in the pathogenesis of autism [30]. The change in the length and grammar of the noncoding region produces eukaryotic speciation and individuality [31]. Changes in the length of noncoding region can lead onto disorders of consciousness like schizophrenia and possibly autism [32]. The integration of nanoarchaea, mevalonate pathway prokaryotes and viroids in to the eukaryotic and human genome produces a chimera which can multiply producing biofilm like multicellular structures having a mixed archaeal, viroidal, prokaryotic and eukaryotic characters which is a regression from the multicellular eukaryotic tissue. This results in a new neuronal, metabolic, immune and tissue phenotype leading to human diseases like autism, cerebral palsy and trisomy 21. The microchimeras formed can lead to polyploidy and trisomy 21.

The archaea and viroids can regulate the nervous system including the NMDA/GABA thalamocorticothalamic pathway mediating conscious perception [2,33]. NMDA/ GABA receptors can be modulated by digoxin induced calcium oscillations resulting NMDA/GAD activity induction, PAH increasing NMDA activity and inducing GAD as well as viroid induced RNA interference [2]. The cholesterol ring oxidase generated pyruvate can be converted by the GABA shunt pathway to glutamate and GABA. The dipolar PAH and archaeal magnetite in the setting of digoxin induced sodium potassium ATPase inhibition can produce a pumped phonon system mediated frohlich model superconducting state inducing quantal perception with nanoarchaeal sensed gravity producing the orchestrated reduction of the quantal possibilities to the macrosopic world [2,33]. The archaea can regulate limbic lobe transmission with archaeal cholesterol aromatase/ring oxidase generated norepinephrine, dopamine, serotonin and acetyl choline [17]. The higher degree of integration of the archaea into the genome produces increased digoxin synthesis producing right hemispheric dominance and lesser degree producing left hemispheric dominance [2]. The increased integration of archaea into the neuronal genome can produce increased cholesterol oxidase and aromatase mediated monoamine and NMDA transmission producing autism, cerebral palsy and trisomy 21. The archaeal bile acids are chemically diverse and structurally different from human bile acids. The archaeal bile acids can bind olfactory GPCR receptors and stimulate the limbic lobe producing a sense of social identity. The dominance of archaeal bile acids over human bile acids in stimulating the olfactory GPCR- limbic lobe pathway leads to loss of social identity and autism [34]. Archaea and RNA viroid can bind the TLR receptor induce NFKB producing immune activation and cytokine TNF alpha secretion. The archaeal DXP and mevalonate pathway metabolites can bind $\gamma\delta$ TCR and digoxin induced calcium signaling can activate NFKB producing chronic immune activation [2,35]. The archaea and viroid induced chronic immune activation and generation of superantigens can lead on to autoimmune disease. Immune activation and autoimmunity is important in the pathogenesis of autism, cerebral palsy and trisomy 21. Archaea, viroids and digoxin can induce the host AKT PI3K, AMPK, HIF alpha and NFKB producing the Warburg metabolic phenotype [36]. The increased glycolytic hexokinase activity, decrease in blood ATP, leakage of cytochrome C, increase in serum pyruvate and decrease in acetyl CoA indicates the generation of the Warburg phenotype. There is induction of glycolysis, inhibition of PDH activity and mitochondrial dysfunction resulting in inefficient energetics. Mitochondrial dysfunction can lead onto NMDA excitotoxicity and cell death important in autism, cerebral palsy and trisomy 21. Cholesterol oxidase activity, increased glycolysis related NADPH oxidase activity and mitochondrial dysfunction generates free radicals important in the pathogenesis of these disease states. The accumulated pyruvate enters the GABA shunt pathway and is converted to citrate which is acted upon by citrate lyase and converted to acetyl CoA, used for cholesterol synthesis [36]. The pyruvate can be converted to glutamate and ammonia which is oxidised by archaea for energy needs. The increased cholesterol substrate leads to increased archaeal growth and digoxin synthesis leading to metabolic channeling to the mevalonate pathway. Hyperdigoxinemia is important in the pathogenesis of trisomy 21 and autism [2]. The Warburg phenotype induced increased mitochondrial PT pore hexokinase, archaeal PAH and viroid induced RNA interference can lead on to malignant transformation important in trisomy 21. The digoxin and PAH induced increased intracellular calcium can lead to PT pore dysfunction, cell death and neuronal degeneration [2]. The archaeal cholesterol catabolism can deplete the cell membranes of cholesterol resulting in organelle dysfunction and neuronal degeneration important in autism, trisomy 21 and cerebral palsy.

References

[1] Valiathan M.S., Somers, K., Kartha, C.C. (1993). *Endomyocardial Fibrosis.* Delhi: Oxford University Press.

[2] Kurup R., Kurup, P.A. (2009). *Hypothalamic digoxin, cerebral dominance and brain function in health and diseases*. New York: Nova Science Publishers.

[3] Hanold D., Randies, J.W. (1991). Coconut cadang-cadang disease and its viroid agent, *Plant Disease,* 75, 330-335.

[4] Edwin B.T., Mohankumaran, C. (2007). Kerala wilt disease phytoplasma: Phylogenetic analysis and identification of a vector, *Proutista moesta, Physiological and Molecular Plant Pathology,* 71(1-3), 41-47.

[5] Eckburg P.B., Lepp, P.W., Relman, D.A. (2003). Archaea and their potential role in human disease, *Infect. Immun.,* 71, 591-596.

[6] Adam Z. (2007). Actinides and Life's Origins, *Astrobiology,* 7, 6-10.

[7] Schoner W. (2002). Endogenous cardiac glycosides, a new class of steroid hormones, *Eur. J. Biochem.,* 269, 2440-2448.

[8] Davies P.C.W., Benner, S.A., Cleland, C.E., Lineweaver, C.H., McKay, C.P., Wolfe-Simon, F. (2009). Signatures of a Shadow Biosphere, *Astrobiology,* 10, 241-249.

[9] Richmond W. (1973). Preparation and properties of a cholesterol oxidase from nocardia species and its application to the enzymatic assay of total cholesterol in serum, *Clin. Chem.,* 19, 1350-1356.

[10] Snell E.D., Snell, C.T. (1961). *Colorimetric Methods of Analysis.* Vol 3A. New York: Van NoStrand.

[11] Glick D. (1971). *Methods of Biochemical Analysis.* Vol 5. New York: Interscience Publishers.

[12] Colowick, Kaplan, N.O. (1955). *Methods in Enzymology.* Vol 2. New York: Academic Press.

[13] Maarten A.H., Marie-Jose, M., Cornelia, G., van Helden–Meewsen, Fritz, E., Marten, P.H. (1995). Detection of muramic acid in human spleen, *Infection and Immunity,* 63(5), 1652 – 1657.

[14] Smit A., Mushegian, A. (2000). Biosynthesis of isoprenoids via mevalonate in Archaea: the lost pathway, *Genome Res,* 10(10), 1468-84.

[15] Van der Geize R., Yam, K., Heuser, T., Wilbrink, M.H., Hara, H., Anderton, M.C. (2007). A gene cluster encoding cholesterol catabolism in a soil actinomycete provides insight into Mycobacterium tuberculosis survival in macrophages, *Proc. Natl. Acad. Sci. USA,* 104(6), 1947-52.

[16] Francis A.J. (1998). Biotransformation of uranium and other actinides in radioactive wastes, *Journal of Alloys and Compounds,* 271(273), 78-84.

[17] Probian C., Wülfing, A., Harder, J. (2003). Anaerobic mineralization of quaternary carbon atoms: Isolation of denitrifying bacteria on pivalic acid (2,2-Dimethylpropionic acid), *Applied and Environmental Microbiology,* 69(3), 1866-1870.

[18] Vainshtein M., Suzina, N., Kudryashova, E., Ariskina, E. (2002). New Magnet-Sensitive Structures in Bacterial and Archaeal Cells, *Biol Cell,* 94(1), 29-35.

[19] Tsagris E.M., de Alba, A.E., Gozmanova, M., Kalantidis, K. (2008). Viroids, *Cell Microbiol,* 10, 2168.

[20] Horie M., Honda, T., Suzuki, Y., Kobayashi, Y., Daito, T., Oshida, T. (2010). Endogenous non-retroviral RNA virus elements in mammalian genomes, *Nature,* 463, 84-87.

[21] Hecht M., Nitz, N., Araujo, P., Sousa, A., Rosa, A., Gomes, D. (2010). Genes from Chagas parasite can transfer to humans and be passed on to children. Inheritance of DNA Transferred from American Trypanosomes to Human Hosts, *PLoS ONE,* 5, 2-10.

[22] Flam F. (1994). Hints of a language in junk DNA, *Science,* 266, 1320.

[23] Horbach S., Sahm, H., Welle, R. (1993). Isoprenoid biosynthesis in bacteria: two different pathways? *FEMS Microbiol Lett,* 111, 135–140.

[24] Gupta R.S. (1998). Protein phylogenetics and signature sequences: a reappraisal of evolutionary relationship among archaebacteria, eubacteria, and eukaryotes, *Microbiol Mol Biol Rev,* 62, 1435–1491.

[25] Hanage W., Fraser, C., Spratt, B. (2005). Fuzzy species among recombinogenic bacteria, *BMC Biology,* 3, 6-10.

[26] Whitchurch C.B., Tolker-Nielsen, T., Ragas, P.C., Mattick, J.S. (2002). Extracellular DNA Required for Bacterial Biofilm Formation, *Science,* 295(5559), 1487.

[27] Webb J.S., Givskov, M., Kjelleberg, S. (2003). Bacterial biofilms: prokaryotic adventures in multicellularity, *Curr. Opin. Microbiol.,* 6(6), 578–85.

[28] Chen Y., Cai, T., Wang, H., Li, Z., Loreaux, E., Lingrel, J.B. (2009). Regulation of intracellular cholesterol distribution by Na/K-ATPase, *J. Biol. Chem.,* 284(22), 14881-90.

[29] Fritzsche M. (2002). Seasonal correlation of sporadic schizophrenia to Ixodes ticks and Lyme borreliosis, *Int. J. Health Geogr,* 1(1), 2.

[30] Waltrip R.W. 2nd, Buchanan, R.W., Summerfelt, A., Breier, A., Carpenter, W.T. Jr., Bryant, N.L., Rubin, S.A., Carbone, K.M. (1995). Borna disease virus and schizophrenia, *Psych. Res.,* 56(1), 33-44.

[31] Poole A.M. (2006). Did group II intron proliferation in an endosymbiont-bearing archaeon create eukaryotes? *Biol. Direct,* 1, 36-40.

[32] Villarreal L.P. (2006). How viruses shape the tree of life, *Future Virology,* 1(5), 587-595.

[33] Lockwood M. (1989). *Mind, Brain and the Quantum*. Oxford: B. Blackwell.

[34] Lefebvre P., Cariou, B., Lien, F., Kuipers, F., Staels, B. (2009). Role of Bile Acids and Bile Acid Receptors in Metabolic Regulation, *Physiol. Rev.,* 89(1), 147 - 191.

[35] Eberl M., Hintz, M., Reichenberg, A., Kollas, A., Wiesner, J., Jomaa, H. (2010). Microbial isoprenoid biosynthesis and human γδ T cell activation, *FEBS Letters,* 544(1), 4-10.

[36] Wallace D.C. (2005). Mitochondria and Cancer: Warburg Addressed, *Cold Spring Harbor Symposia on Quantitative Biology,* 70, 363-374.

Chapter VII

Actinide Dependent Shadow Biosphere of Archaea and Viroids and Hemispheric Dominance

Abbreviation

BHD: Bihemispheric dominance
RHD: Right hemispheric dominance
LHD: Left hemispheric dominance

Introduction

The human brain synthesises an endogenous membrane sodium-potassium ATPase inhibitor digoxin which plays a role in neuro-immuno-endocrine integration and pathogenesis of several neuropsychiatric and systemic diseases. Endomyocardial fibrosis (EMF) along with the root wilt disease of coconut is endemic to Kerala with its radioactive actinide beach sands. Actinides like rutile producing intracellular magnesium deficiency due to rutile-magnesium exchange sites in the cell membrane has been implicated in the etiology of EMF [1]. Endogenous digoxin, a steroidal glycoside which functions as a membrane sodium-potassium ATPase inhibitor has also been related to its etiology due to the intracellular magnesium deficiency it produces [2]. Organisms like phytoplasmas and viroids have also been demonstrated to play a role in the etiology of these diseases [3,4]. Endogenous digoxin has been related to hemispheric dominance [2]. Right hemispheric dominant individuals were hyperdigoxinemic, left hemispheric dominant individuals were hypodigoxinemic and bihemispheric dominant individuals were normodigoxinemic. The possibility of endogenous digoxin synthesis by actinide based primitive organism like archaea with a mevalonate pathway and cholesterol catabolism was considered [5,6,7]. Davies has put forward the concept of a shadow biosphere of organisms with alternate biochemistry present in earth itself [8]. An actinide dependent shadow biosphere of archaea and viroids in the above mentioned disease states is described [6]. The intracellular endosymbionts archaea and their intron derived viroids constitute the third element regulating the human body.

Methods

Informed consent of the subjects and the approval of the ethics committee were obtained for the study. The following groups were included in the study:- (I) right handed and left hemispheric dominant group, (II) left handed and right hemispheric dominant group and (III) amphidextrous and bihemispheric dominant individuals. Hemispheric dominance was assessed by methods described in previous reports [2]. There were 10 healthy normal individuals in the age range between 20 to 30 years in each group. They were selected randomly from the general population. The blood samples were drawn in the fasting state. Plasma from fasting heparinised blood was used and the experimental protocol was as follows (I) Plasma+phosphate buffered saline, (II) same as I+cholesterol substrate, (III) same as II+rutile 0.1 mg/ml, (IV) same as II+ciprofloxacine and doxycycline each in a concentration of 1 mg/ml. Cholesterol substrate was prepared as described by Richmond [9]. Aliquots were withdrawn at zero time immediately after mixing and after incubation at 37°C for 1 hour. The following estimations were carried out:- Cytochrome F420, free RNA, free DNA, muramic acid, polycyclic aromatic hydrocarbon, hydrogen peroxide, dopamine, pyruvate, ammonia, glutamate, cytochrome C, hexokinase, ATP synthase, HMG CoA reductase, digoxin and bile acids [10,11,12,13]. Cytochrome F420 was estimated flourimetrically (excitation wavelength 420 nm and emission wavelength 520 nm). Polycyclic aromatic hydrocarbon was estimated by measuring hydrogen peroxide liberated by using glucose reagent. The statistical analysis was done by ANOVA.

Results

The parameters checked as indicated above were:- cytochrome F420, free RNA, free DNA, muramic acid, polycyclic aromatic hydrocarbon, hydrogen peroxide, serotonin, pyruvate, ammonia, glutamate, cytochrome C, hexokinase, ATP synthase, HMG CoA reductase, digoxin and bile acids. The plasma of the bihemispheric dominant group showed detectable levels of the above mentioned parameters after incubation for 1 hour and addition of cholesterol substrate resulted in still further increase in these parameters. The addition of antibiotics to the bihemispheric dominant group caused a decrease in all the parameters while addition of rutile increased their levels. The plasma of right hemispheric dominant group showed a significant increase in the above mentioned parameters as compared to bihemispheric dominance group. The addition of antibiotics to the right hemispheric dominant group caused a decrease in all the parameters while addition of rutile increased their levels but the extent of change was more in right hemispheric dominant group as compared to bihemispheric dominant group. The plasma of left hemispheric dominant group showed a significant decrease in the above mentioned parameters as compared to the bihemispheric dominant group. The addition of antibiotics to the left hemispheric dominant group caused a decrease in all the parameters while addition of rutile increased their levels but the extent of change was less in left hemispheric dominant group as compared to bihemispheric dominant group. The results are expressed in tables 1-7 as percentage change in the parameters after 1 hour incubation as compared to the values at zero time.

Discussion

There was increase in cytochrome F420 indicating archaeal growth. The archaea can synthesise and use cholesterol as a carbon and energy source [14,15]. The archaeal origin of the enzyme activities was indicated by antibiotic induced suppression. The study indicates the presence of actinide based archaea with an alternate actinide based enzymes or metalloenzymes in the system as indicated by rutile induced increase in enzyme activities [16]. There was also an increase in archaeal HMG CoA reductase activity indicating increased cholesterol synthesis by the archaeal mevalonate pathway.

The archaeal beta hydroxyl steroid dehydrogenase activity indicating digoxin synthesis and archaeal cholesterol hydroxylase activity indicating bile acid synthesis were increased [7].

The archaeal cholesterol oxidase activity was increased resulting in generation of pyruvate and hydrogen peroxide [15]. The pyruvate gets converted to glutamate and ammonia by the GABA shunt pathway. The archaeal aromatization of cholesterol generating PAH, serotonin and dopamine was also detected [17]. The archaeal glycolytic hexokinase activity and archaeal extracellular ATP synthase activity were increased. The archaea can undergo magnetite and calcium carbonate mineralization and can exist as calcified nanoforms [18]. There was an increase in free RNA indicating self replicating RNA viroids and free DNA indicating generation of viroid complementary DNA strands by archaeal reverse transcriptase activity. The actinides modulate RNA folding and catalyse its ribozymal action. Digoxin can cut and paste the viroidal strands by modulating RNA splicing generating RNA viroidal diversity.

The viroids are evolutionarily escaped archaeal group I introns which have retrotransposition and self splicing qualities [19]. The decrease in free self replicating RNA and DNA with the addition of antibiotics indicates that the RNA viroids are derived from archaeal introns. Archaeal pyruvate can produce histone deacetylase inhibition resulting in endogenous retroviral (HERV) reverse transcriptase and integrase expression. This can integrate the RNA viroidal complementary DNA into the noncoding region of eukaryotic non coding DNA using HERV integrase as has been described for borna and ebola viruses [20]. The noncoding DNA is lengthened by integrating RNA viroidal complementary DNA with the integration going on as a continuing event.

The archaea genome can also get integrated into human genome using integrase as has been described for trypanosomes [21]. The integrated viroids and archaea can undergo vertical transmission and can exist as genomic parasites [20,21]. This increases the length and alters the grammar of the noncoding region producing memes or memory of acquired characters [22]. The viroidal complementary DNA can function as jumping genes producing a dynamic genome important in storage of synaptic information, HLA gene expression and neurodevelopmental gene expression.

Table 1. Effect of rutile and antibiotics on free DNA and RNA

Group	DNA % change (Increase with Rutile)		DNA % change (Decrease with antibiotics)		RNA % change (Increase with Rutile)		RNA % change (Decrease with antibiotics)	
	Mean	± SD	Mean	± SD	Mean	± SD	Mean	± SD
BHD	4.37	0.15	18.39	0.38	4.37	0.13	18.38	0.48
RHD	22.99	1.56	65.19	4.10	23.27	1.36	65.66	3.93
LHD	2.26	0.25	7.45	0.40	2.30	0.12	7.62	0.30
F value	337.577		356.621		427.828		654.453	
P value	< 0.001		< 0.001		< 0.001		< 0.001	

Table 2. Effect of rutile and antibiotics on cyt F 420 and muramic acid

Group	CYT F420 % change (Increase with Rutile)		CYT F420 % change (Decrease with antibiotics)		Muramic acid % change (Increase with Rutile)		Muramic acid % change (Decrease with antibiotics)	
	Mean	± SD	Mean	± SD	Mean	± SD	Mean	± SD
BHD	4.48	0.15	18.24	0.66	4.34	0.15	18.24	0.37
RHD	11.35	0.64	60.49	6.22	22.68	1.99	63.29	5.93
LHD	2.13	0.13	5.37	1.47	2.26	0.25	7.45	0.40
F value	306.749		130.054		348.867		364.999	
P value	< 0.001		< 0.001		< 0.001		< 0.001	

Table 3. Effect of rutile and antibiotics on HMG CoA reductase and PAH

Group	HMG CoA R % change (Increase with Rutile)		HMG CoA R % change (Decrease with antibiotics)		PAH % change (Increase with Rutile)		PAH % change (Decrease with antibiotics)	
	Mean	± SD	Mean	± SD	Mean	± SD	Mean	± SD
BHD	4.30	0.20	18.35	0.35	4.45	0.14	18.25	0.72
RHD	21.06	2.32	63.87	6.22	21.00	2.54	57.42	7.07
LHD	2.33	0.17	7.24	0.59	2.25	0.17	7.01	0.65
F value	319.332		199.553		391.318		257.996	
P value	< 0.001		< 0.001		< 0.001		< 0.001	

Table 4. Effect of rutile and antibiotics on digoxin and bile acids

Group	Digoxin (ng/ml) (Increase with Rutile)		Digoxin (ng/ml) (Decrease with antibiotics)		Bile Acids % change (Increase with Rutile)		Bile Acids % change (Decrease with antibiotics)	
	Mean	± SD	Mean	± SD	Mean	± SD	Mean	± SD
BHD	0.11	0.00	0.054	0.003	4.29	0.18	18.15	0.58
RHD	0.55	0.10	0.248	0.058	21.10	2.43	54.82	8.28
LHD	0.07	0.01	0.026	0.004	2.25	0.19	7.25	0.66
F value	135.116		7[illegible].706		290.441		203.651	
P value	< 0.001		< 0.001		< 0.001		< 0.001	

Table 5. Effect of rutile and antibiotics on pyruvate and hexokinase

Group	Pyruvate % change (Increase with Rutile)		Pyruvate % change (Decrease with antibiotics)		Hexokinase % change (Increase with Rutile)		Hexokinase % change (Decrease with antibiotics)	
	Mean	± SD	Mean	± SD	Mean	± SD	Mean	± SD
BHD	4.34	0.21	18.43	0.82	4.21	0.16	18.56	0.76
RHD	11.12	0.66	59.68	6.24	23.27	1.68	67.35	3.77
LHD	2.16	0.18	5.91	1.38	2.24	0.17	6.29	1.06
F value	321.255		115.242		292.065		317.966	
P value	< 0.001		< 0.001		< 0.001		< 0.001	

Table 6. Effect of rutile and antibiotics on ATP synthase and hydrogen peroxide

Group	ATP synthase % change (Increase with Rutile)		ATP synthase % change (Decrease with antibiotics)		H_2O_2 % change (Increase with Rutile)		H_2O_2 % change (Decrease with antibiotics)	
	Mean	± SD	Mean	± SD	Mean	± SD	Mean	± SD
BHD	4.40	0.11	18.78	0.11	4.43	0.19	18.13	0.63
RHD	11.99	0.38	56.34	3.39	17.60	3.53	54.68	5.09
LHD	2.30	0.12	7.62	0.30	2.24	0.23	5.36	0.99
F value	449.503		573.081		380.721		171.228	
P value	< 0.001		< 0.001		< 0.001		< 0.001	

Table 7. Effect of rutile and antibiotics on delta amino levulinic acid and dopamine

Group	ALA % (Increase with Rutile)		ALA % (Decrease with antibiotics)		DOPAMINE % change (Increase with Rutile)		DOPAMINE % change (Decrease with antibiotics)	
	Mean	± SD	Mean	± SD	Mean	± SD	Mean	± SD
BHD	4.40	0.10	18.48	0.39	4.41	0.15	18.63	0.12
RHD	22.98	2.06	66.10	4.03	11.36	0.58	65.41	4.83
LHD	2.13	0.11	7.62	0.32	2.13	0.11	7.62	0.32
F value	372.716		556.411		403.394		680.284	
P value	< 0.001		< 0.001		< 0.001		< 0.001	

The alteration in DNA sequences produced by viroidal complementary DNA jumping genes can lead onto schizophrenia and primary seizure disorder. The RNA viroids can regulate mRNA function by RNA interference [19]. The phenomena of RNA interference can modulate T cell and B cell function, neuronal transmission and euchromatin/heterochromatin expression. The RNA viroid induced mRNA interference can modulate dopaminergic, glutamatergic and serotoninergic synaptic transmission. The archaea and viroidal density is high in right hemispheric dominance, intermediate in bihemispheric dominance and low in left hemispheric dominance.

The presence of muramic acid, HMG CoA reductase and cholesterol oxidase activity inhibited by antibiotics indicates the presence of bacteria with mevalonate pathway. The density of the mevalonate pathway bacterial is high in right hemispheric dominance, low in left hemispheric dominance and intermediate in bihemispheric dominance. The bacterial with mevalonate pathway include streptococcus, staphylococcus, actinomycetes, listeria, coxiella and borrelia [23]. The bacteria and archaea with mevalonate pathway and cholesterol catabolism had a evolutionarily advantage and constitutes the isoprenoidal clade organism with the archaea evolving into mevalonate pathway gram positive and gram negative organism through horizontal gene transfer of viroidal and virus genes [24]. The isoprenoidal clade prokaryotes develop into other groups of prokaryotes via viroidal/virus as well as eukaryotic horizontal gene transfer producing bacterial speciation [25].

The RNA viroids and its complementary DNA developed into cholesterol enveloped RNA and DNA viruses like herpes, retrovirus, influenza virus, borna virus, cytomegalo virus and Ebstein Barr virus by recombining with eukaryotic and human genes resulting in viral speciation. Bacterial and viral species are ill defined and fuzzy with all of them forming one common genetic pool with frequent horizontal gene transfer and recombination. Thus the multi and unicellular eukaryote with its genes serves the purpose of prokaryotic and viral speciation .The multicellular eukaryote developed so that their endosymbiotic archaeal colonies could survive and forage better. The multicellular eukaryotes are like bacterial biofilms. The archaea and bacteria with a mevalonate pathway uses the extracellular RNA viroids and DNA viroids for quorum sensing and in the generation of symbiotic biofilm like structures which develop into multicellular eukaryotes [26,27]. The endosymbiotic archaea and bacteria with mevalonate pathway still uses the RNA viroids and DNA viroids for the regulation of muticellular eukaryote. Pollution is induced by the primitive nanoarchaea and mevalonate pathway bacteria synthesised PAH and methane leading on to redox stress. Redox stress leads to sodium potassium ATPase inhibition, inward movement of plasma membrane cholesterol, defective SREBP sensing, increased cholesterol synthesis and nanoarchaeal/ mevalonate pathway bacterial growth [28]. Redox stress leads on to viroidal and archaeal multiplication. Redox stress can also lead to HERV reverse transcriptase and integrase expression. The noncoding DNA is formed of integrating RNA viroidal complementary DNA and archaea with the integration going on as a continuing event. The archaeal pox like dsDNA virus forms evolutionarily the nucleus. The integrated viroidal, archaeal and mevalonate pathway bacterial sequences can undergo vertical transmission and can exist as genomic parasites. The genomic integrated archaea, mevalonate pathway bacteria and viroids form a genomic reserve of bacteria and viruses which can recombine with human and eukaryotic genes producing bacterial and viral speciation. Bacteria and viruses can contribute to the regulation of hemispheric dominance as exemplified by schizophrenia, a disorder of consciousness. Borrelia, toxoplasma, chlamydia, mycoplasma, retroviruses, herpes virus, influenza virus and borna virus contribute to the neuropathogenesis of schizophrenia [29,30,31]. The change in the length and grammar of the noncoding region produces eukaryotic speciation and individuality [32]. Changes in the length of noncoding region can lead onto modulation of hemispheric dominance and conscious perception as exemplified in schizophrenia [33]. The human endogenous retroviruses and change in the length and grammar of the noncoding region has been described in schizophrenia. The integration of nanoarchaea, mevalonate pathway prokaryotes and viroids in to the eukaryotic and human genome produces a chimera which can multiply producing biofilm like multicellular structures having a mixed archaeal, viroidal, prokaryotic and eukaryotic characters which is a regression from the multicellular eukaryotic tissue. This results in a new neuronal, metabolic, immune and tissue phenotype producing microchimeras. Microchimeras can also generate tissue and neuronal polyploidy. The higher degree of integration of archaea, mevalonate pathway bacteria and viorids into the genome produces right hemispheric dominance, intermediate degree of integration produces bihemispheric dominance and lower degree of integration left hemispheric dominance.

The archaea and viroids can regulate the nervous system including the NMDA/GABA thalamocorticothalamic pathway mediating conscious perception [2,34]. NMDA/GABA receptors can be modulated by digoxin induced calcium oscillations resulting in NMDA/glutamic acid decarboxylase (GAD) activity induction, PAH increasing NMDA activity and inducing GAD as well as viroid induced RNA interference modulating NMDA/GABA receptors[2]. The cholesterol ring oxidase generated pyruvate can be converted by the GABA shunt pathway to glutamate and GABA. Increased NMDA transmission has been described in schizophrenia and primary seizure disorder. The dipolar PAH and archaeal magnetite in the setting of digoxin induced sodium potassium ATPase inhibition can produce a pumped phonon system mediated frohlich model superconducting state inducing quantal perception with nanoarchaeal sensed gravity producing the orchestrated reduction of the quantal possibilities to the macrosopic world [2,34]. The archaea can regulate limbic lobe transmission with archaeal cholesterol aromatase/ring oxidase generated norepinephrine, dopamine, serotonin and acetyl choline [17]. Thus the shadow biosphere of archaea and viroids can regulate conscious and quantal perception. The archaea and viriods can also modulate multiple neurotransmitter systems. Schizophrenia is described as a disorder of consciousness and increased integration of archaea and viroids into the genome can contribute to its neuropathogenesis. Increased dopaminergic, serotoninergic and NMDA transmission is important in the pathogenesis of schizophrenia. The higher degree of integration of the archaea into the genome produces increased digoxin synthesis producing right hemispheric dominance and lesser degree producing left hemispheric dominance [2]. Bihemispheric dominance is intermediate with normal digoxin synthesis. Right hemispheric dominance has been described in schizophrenia. The increased integration of archaea into the neuronal genome can produce increased cholesterol oxidase and aromatase mediated monoamine and NMDA transmission producing schizophrenia. The archaeal bile acids are chemically diverse and structurally different from human bile acids. The archaeal bile acids can bind olfactory GPCR receptors and stimulate the limbic lobe producing a sense of social identity. The dominance of archaeal bile acids over human bile acids in stimulating the olfactory GPCR-limbic lobe pathway leads to loss of social identity leading to schizophrenia and autism [35].

Archaea and RNA viroid can bind the TLR receptor induce NFKB producing immune activation and cytokine TNF alpha secretion. The archaeal DXP and mevalonate pathway metabolites can bind γδ TCR and digoxin induced calcium signaling can activate NFKB producing chronic immune activation [2,36]. The archaea and viroid induced chronic immune activation and generation of superantigens can lead on to autoimmune disease. This produces a state of chronic immune activation in right hemispheric dominance producing increased predisposition to autoimmune diseases. The left hemispheric dominant group is immunosuppressed and the bihemispheric dominant group has normal immune function.

Archaea, viroids and digoxin can induce the host AKT PI3K, AMPK, HIF alpha and NFKB producing the Warburg metabolic phenotype [37]. The increased glycolytic hexokinase activity, decrease in blood ATP, leakage of cytochrome C, increase in serum pyruvate and decrease in acetyl CoA indicates the generation of the Warburg phenotype. There is induction of glycolysis, inhibition of PDH activity and mitochondrial dysfunction resulting in inefficient energetics. The immune activation mediated increased levels of TNF alpha can produce insulin resistance acting at the level of insulin receptor. Thus a state similar to metabolic syndrome X exists in right hemispheric dominance. Left hemispheric dominance can have a pattern of insulin sensitivity while bihemispheric dominance will be metabolically intermediate. Cholesterol oxidase activity, increased glycolysis related NADPH oxidase activity and mitochondrial dysfunction generates free radicals. Free radical production and mitochondrial dysfunction can increase NMDA transmission important in conscious perception. The accumulated pyruvate enters the gaba shunt pathway and is converted to citrate which is acted upon by citrate lyase and converted to acetyl CoA, used for cholesterol synthesis [37]. The increased cholesterol substrate leads to increased archaeal growth and digoxin synthesis leading to metabolic channeling to the mevalonate pathway. Hyperdigoxinemia is important in the regulation of hemispheric dominance [2]. The right hemispheric dominant group is hyperdigoxinemic, left hemispheric dominant group is hypodigoxinemic and bihemispheric dominant group is normodigoxinemic. The pyruvate can be converted to glutamate and ammonia which is oxidised by archaea for energy needs. Ammonia can regulate both NMDA and GABA transmission depending on its levels. The Warburg phenotype can contribute to the hemispheric dominance by augmenting the bacterial shikimic acid pathway. The upregulated glycolysis consequent to the Warburg phenotype produces phosphoenolpyruvate, a basic substrate for the bacterial shikimic acid pathway which can synthesise monoamines and neuroactive alkaloids. The shikimic acid pathway can generate dopamine and serotonin producing increased monoaminergic transmission. The shikimic acid pathway can also synthesise the neuroactive alkaloids strychnine, nicotine, morphine, mescaline and LSD important in regulating neural transmission [2]. The upregulated glycolysis can also contribute to increased NMDA and GABA transmission in the thalamocorticothalamic pathway. The glycolytic pathway produces phosphoglycerate which is converted to phosphoserine and then serine which activates the NMDA receptor. The glycolytic enzyme glyceraldehyde-3-phosphate dehydrogenase is a GABA receptor kinase and activates GABA transmission. Thus the archaea and viroid induced Warburg phenotype can contribute to the modulation of hemispheric dominance by regulating the multiple neurotransmitter systems. The archaeal cholesterol catabolism can deplete the cell membranes of cholesterol resulting in alteration in lipid microdomains and their related neurotransmitter receptor contributing to the regulation of NMDA, serotoninergic and dopaminergic transmission. Thus the archaeal cholesterol catabolism and viroids can regulate brain function and hemispheric dominance. The archaea and viroids have axonal and transynaptic transport functioning as biological neurotransmitters. The brain can be visualized evolutionarily as a modified mevalonate pathway bacteria and archaeal colony functioning by mechanisms of quorum sensing using RNA viroids with its bacterial flagellar system forming axo-axonic and axo-dendritic connections. The third element of archaea and their derived viroids can also regulate the immune, genetic, metabolic and neural systems producing its integration.

The third element formed of intracellular archaea and viroidal symbiosis determines hemispheric dominance. Archaeal cholesterol synthesis and catabolism determines hemispheric dominance.

References

[1] Valiathan M.S., Somers, K., Kartha, C.C. (1993). *Endomyocardial Fibrosis.* Delhi: Oxford University Press.

[2] Kurup R., Kurup, P.A. (2009). *Hypothalamic digoxin, cerebral dominance and brain function in health and diseases.* New York: Nova Science Publishers.

[3] Hanold D., Randies, J.W. (1991). Coconut cadang-cadang disease and its viroid agent, *Plant Disease,* 75, 330-335.

[4] Edwin B.T., Mohankumaran, C. (2007). Kerala wilt disease phytoplasma: Phylogenetic analysis and identification of a vector, *Proutista moesta, Physiological and Molecular Plant Pathology,* 71(1-3), 41-47.

[5] Eckburg P.B., Lepp, P.W., Relman, D.A. (2003). Archaea and their potential role in human disease, *Infect Immun,* 71, 591-596.

[6] Adam Z. (2007). Actinides and Life's Origins, *Astrobiology,* 7, 6-10.

[7] Schoner W. (2002). Endogenous cardiac glycosides, a new class of steroid hormones, *Eur. J. Biochem.,* 269, 2440-2448.

[8] Davies P.C.W., Benner, S.A., Cleland, C.E., Lineweaver, C.H., McKay, C.P., Wolfe-Simon, F. (2009). Signatures of a Shadow Biosphere, *Astrobiology,* 10, 241-249.

[9] Richmond W. (1973). Preparation and properties of a cholesterol oxidase from nocardia species and its application to the enzymatic assay of total cholesterol in serum, *Clin. Chem.,* 19, 1350-1356.

[10] Snell E.D., Snell, C.T. (1961). *Colorimetric Methods of Analysis.* Vol 3A. New York: Van NoStrand.

[11] Glick D. (1971). *Methods of Biochemical Analysis.* Vol 5. New York: Interscience Publishers.

[12] Colowick, Kaplan, N.O. (1955). *Methods in Enzymology.* Vol 2. New York: Academic Press.

[13] Maarten A.H., Marie-Jose, M., Cornelia, G., van Helden–Meewsen, Fritz, E., Marten, P.H. (1995). Detection of muramic acid in human spleen, *Infection and Immunity,* 63(5), 1652 – 1657.

[14] Smit A., Mushegian, A. (2000). Biosynthesis of isoprenoids via mevalonate in Archaea: the lost pathway, *Genome Res.,* 10(10), 1468-84.

[15] Van der Geize R., Yam, K., Heuser, T., Wilbrink, M.H., Hara, H., Anderton, M.C. (2007). A gene cluster encoding cholesterol catabolism in a soil actinomycete provides insight into Mycobacterium tuberculosis survival in macrophages, *Proc. Natl. Acad. Sci. USA,* 104(6), 1947-52.

[16] Francis A.J. (1998). Biotransformation of uranium and other actinides in radioactive wastes, *Journal of Alloys and Compounds,* 271(273), 78-84.

[17] Probian C., Wülfing, A., Harder, J. (2003). Anaerobic mineralization of quaternary carbon atoms: Isolation of denitrifying bacteria on pivalic acid (2,2-Dimethylpropionic acid), *Applied and Environmental Microbiology,* 69(3), 1866-1870.

[18] Vainshtein M., Suzina, N., Kudryashova, E., Ariskina, E. (2002). New Magnet-Sensitive Structures in Bacterial and Archaeal Cells, *Biol. Cell,* 94(1), 29-35.

[19] Tsagris E.M., de Alba, A.E., Gozmanova, M., Kalantidis, K. (2008). Viroids, *Cell Microbiol.,* 10, 2168.

[20] Horie M., Honda, T., Suzuki, Y., Kobayashi, Y., Daito, T., Oshida, T. (2010). Endogenous non-retroviral RNA virus elements in mammalian genomes, *Nature,* 463, 84-87.

[21] Hecht M., Nitz, N., Araujo, P., Sousa, A., Rosa, A., Gomes, D. (2010). Genes from Chagas parasite can transfer to humans and be passed on to children. Inheritance of DNA Transferred from American Trypanosomes to Human Hosts, *PLoS ONE,* 5, 2-10.

[22] Flam F. (1994). Hints of a language in junk DNA, *Science,* 266, 1320.

[23] Horbach S., Sahm, H., Welle, R. (1993). Isoprenoid biosynthesis in bacteria: two different pathways? *FEMS Microbiol. Lett.,* 111, 135–140.

[24] Gupta R.S. (1998). Protein phylogenetics and signature sequences: a reappraisal of evolutionary relationship among archaebacteria, eubacteria, and eukaryotes, *Microbiol. Mol. Biol. Rev.,* 62, 1435–1491.

[25] Hanage W., Fraser, C., Spratt, B. (2005). Fuzzy species among recombinogenic bacteria, *BMC Biology,* 3, 6-10.

[26] Webb J.S., Givskov, M., Kjelleberg, S. (2003). Bacterial biofilms: prokaryotic adventures in multicellularity, *Curr. Opin. Microbiol.,* 6(6), 578–85.

[27] Whitchurch C.B., Tolker-Nielsen, T., Ragas, P.C., Mattick, J.S. (2002). Extracellular DNA Required for Bacterial Biofilm Formation. *Science,* 295(5559), 1487.

[28] Chen Y., Cai, T., Wang, H., Li, Z., Loreaux, E., Lingrel, J.B. (2009). Regulation of intracellular cholesterol distribution by Na/K-ATPase, *J. Biol. Chem.,* 284(22), 14881-90.

[29] Fritzsche M. (2002). Seasonal correlation of sporadic schizophrenia to Ixodes ticks and Lyme borreliosis. *Int. J. Health Geogr,* 1(1), 2.

[30] Waltrip R.W. 2nd, Buchanan, R.W., Summerfelt, A., Breier, A., Carpenter, W.T. Jr., Bryant, N.L., Rubin, S.A., Carbone, K.M. (1995). Borna disease virus and schizophrenia. *Psych. Res.,* 56(1), 33-44.

[31] Torrey E.F., Yolken, R.H. (2003). Toxoplasma gondii and schizophrenia. *Emerg. Infect. Dis.,* 9(11), 1375-80.

[32] Poole A.M. (2006). Did group II intron proliferation in an endosymbiont-bearing archaeon create eukaryotes? *Biol. Direct,* 1, 36-40.

[33] Villarreal L.P. (2006). How viruses shape the tree of life, *Future Virology,* 1(5), 587-595.

[34] Lockwood M. (1989). *Mind, Brain and the Quantum*. Oxford: B. Blackwell.

[35] Lefebvre P., Cariou, B., Lien, F., Kuipers, F., Staels, B. (2009). Role of Bile Acids and Bile Acid Receptors in Metabolic Regulation, *Physiol Rev,* 89(1), 147 - 191.

[36] Eberl M., Hintz, M., Reichenberg, A., Kollas, A., Wiesner, J., Jomaa, H. (2010). Microbial isoprenoid biosynthesis and human γδ T cell activation, *FEBS Letters,* 544(1), 4-10.

[37] Wallace D.C. (2005). Mitochondria and Cancer: Warburg Addressed, *Cold Spring Harbor Symposia on Quantitative Biology,* 70, 363-374.

Chapter VIII

A Cholesterol and Actinide Dependent Shadow Biosphere of Archaea and Viroids in Pulmonary Diseases

Introduction

Endomyocardial fibrosis (EMF) along with the root wilt disease of coconut is endemic to Kerala with its radioactive actinide beach sands. Actinides like rutile producing intracellular magnesium deficiency due to rutile-magnesium exchange sites in the cell membrane has been implicated in the etiology of EMF.[1] Endogenous digoxin, a steroidal glycoside which functions as a membrane sodium-potassium ATPase inhibitor has also been related to its etiology due to the intracellular magnesium deficiency it produces.[2] Organisms like phytoplasmas and viroids have also been demonstrated to play a role in the etiology of these diseases.[3,4] Endogenous digoxin has been related to the pathogenesis of interstitial lung disease, chronic bronchitis emphysema and bronchial asthma.[2] The possibility of endogenous digoxin synthesis by actinide based primitive organism like archaea with a mevalonate pathway and cholesterol catabolism was considered.[5,6,7] Davies has put forward the concept of a shadow biosphere of organisms with alternate biochemistry present in earth itself.[8] An actinide dependent shadow biosphere of archaea and viroids in the above mentioned disease states is described [6].

Methods

Informed consent of the subjects and the approval of the ethics committee were obtained for the study. The following groups were included in the study:- interstitial lung disease, chronic bronchitis emphysema and bronchial asthma. There were 10 patients in each group and each patient had an age and sex matched healthy control selected randomly from the general population. The blood samples were drawn in the fasting state before treatment was initiated. Plasma from fasting heparinised blood was used and the experimental protocol was as follows (I) Plasma+phosphate buffered saline, (II) same as I+cholesterol substrate, (III) same as II+rutile 0.1 mg/ml, (IV) same as II+ciprofloxacine and doxycycline each in a concentration of 1 mg/ml. Cholesterol substrate was prepared as described by Richmond. [9] Aliquots were withdrawn at zero time immediately after mixing and after incubation at 37°C for 1 hour. The following estimations were carried out:- Cytochrome F420, free RNA, free DNA, muramic acid, polycyclic aromatic hydrocarbon, hydrogen peroxide, serotonin, pyruvate, ammonia, glutamate, cytochrome C, hexokinase, ATP synthase, HMG CoA redutase, digoxin and bile acids.[10,11,12,13] Cytochrome F420 was estimated flourimetrically (excitation wavelength 420 nm and emission wavelength 520 nm). Polycyclic aromatic hydrocarbon was estimated by measuring hydrogen peroxide liberated by using glucose reagent. The statistical analysis was done by ANOVA.

Results

The parameters checked as indicated above were:- cytochrome F420, free RNA, free DNA, muramic acid, polycyclic aromatic hydrocarbon, hydrogen peroxide, serotonin, pyruvate, ammonia, glutamate, cytochrome C, hexokinase, ATP synthase, HMG CoA redutase, digoxin and bile acids. Plasma of control subjects showed increased levels of the above mentioned parameters with after incubation for 1 hour and addition of cholesterol substrate resulted in still further significant increase in these parameters. The plasma of patients showed similar results but the extent of increase was more. The addition of antibiotics to the control plasma caused a decrease in all the parameters while addition of rutile increased their levels. The addition of antibiotics to the patient's plasma caused a decrease in all the parameters while addition of rutile increased their levels but the extent of change was more in patient's sera as compared to controls. The results are expressed in tables 1-7 as percentage change in the parameters after 1 hour incubation as compared to the values at zero time.

Discussion

There was increase in cytochrome F420 indicating archaeal growth in interstitial lung disease, chronic bronchitis emphysema and bronchial asthma. The archaea can synthesise and use cholesterol as a carbon and energy source.[14,15] The archaeal origin of the enzyme activities was indicated by antibiotic induced suppression.

The study indicates the presence of actinide based archaea with an alternate actinide based enzymes or metalloenzymes in the system as indicated by rutile induced increase in enzyme activities [16]. There was also an increase in archaeal HMG CoA reductase activity indicating increased cholesterol synthesis by the archaeal mevalonate pathway. The archaeal beta hydroxyl steroid dehydrogenase activity indicating digoxin synthesis and archaeal cholesterol hydroxylase activity indicating bile acid synthesis were increased.[7] The archaeal cholesterol oxidase activity was increased resulting in generation of pyruvate and hydrogen peroxide.[15] The pyruvate gets converted to glutamate and ammonia by the GABA shunt pathway. The archaeal aromatization of cholesterol generating PAH, serotonin and dopamine was also detected [17].

Table 1. Effect of rutile and antibiotics on muramic acid and serotonin

	Muramic acid % change (Increase with Rutile)		Muramic acid % change (Decrease with Doxy+Cipro)		5 HT % (Increase without Doxy)		5 HT % (Decrease with Doxy)	
Group	Mean	± SD	Mean	± SD	Mean	± SD	Mean	± SD
Normal	4.41	0.15	18.63	0.12	4.34	0.15	18.24	0.37
ASTH	23.45	1.79	66.32	3.63	22.56	2.46	62.70	4.53
CBE	23.20	1.57	66.65	4.26	22.12	2.44	63.69	5.14
ILD	22.95	1.61	65.76	4.01	22.92	1.99	66.55	4.55
F value	403.394		680.284		348.867		364.999	
P value	< 0.001		< 0.001		< 0.001		< 0.001	

Table 2. Effect of rutile and antibiotics on free DNA and RNA

	DNA % change (Increase with Rutile)		DNA % change (Decrease with Doxy)		RNA % change (Increase with Rutile)		RNA % change (Decrease with Doxy)	
Normal	4.37	0.15	18.39	0.38	4.37	0.13	18.38	0.48
ASTH	23.17	1.49	63.96	5.72	23.21	1.72	66.40	3.69
CBE	22.98	1.50	65.13	4.87	23.15	1.62	66.48	4.17
ILD	22.79	2.20	64.26	6.02	22.60	1.64	66.86	4.21
F value	337.577		356.621		427.828		654.453	
P value	< 0.001		< 0.001		< 0.001		< 0.001	

Table 3. Effect of rutile and antibiotics on HMG CoA reductase and PAH

	HMG CoA R % change (Increase with Rutile)		HMG CoA R % change (Decrease with Doxy)		PAH % change (Increase with Rutile)		PAH % change (Decrease with Doxy)	
Normal	4.30	0.20	18.35	0.35	4.45	0.14	18.25	0.72
ASTH	22.03	2.58	60.73	5.55	23.22	1.67	62.06	2.05
CBE	22.28	2.10	65.21	3.81	23.31	1.70	64.38	5.67
ILD	22.75	2.75	62.71	6.04	22.98	2.01	63.91	4.86
F value	319.332		199.553		391.318		257.996	
P value	< 0.001		< 0.001		< 0.001		< 0.001	

Table 4. Effect of rutile and antibiotics on digoxin and bile acids

	Digoxin (ng/ml) (Increase with Rutile)		Digoxin (ng/ml) (Decrease with Doxy+Cipro)		Bile acids % change (Increase with Rutile)		Bile acids % change (Decrease with Doxy)	
Normal	0.11	0.00	0.054	0.003	4.29	0.18	18.15	0.58
ASTH	0.49	0.07	0.199	0.022	23.38	2.13	64.52	6.49
CBE	0.55	0.04	0.219	0.038	23.19	1.72	64.25	6.19
ILD	0.50	0.07	0.183	0.029	22.77	1.97	64.79	5.78
F value	135.116		71.706		290.441		203.651	
P value	< 0.001		< 0.001		< 0.001		< 0.001	

Table 5. Effect of rutile and antibiotics on pyruvate and hexokinase

	Pyruvate % change (Increase with Rutile)		Pyruvate % change (Decrease with Doxy)		Hexokinase % change (Increase with Rutile)		Hexokinase % change (Decrease with Doxy)	
Normal	4.34	0.21	18.43	0.82	4.21	0.16	18.56	0.76
ASTH	21.26	2.04	55.44	7.92	22.20	2.41	64.44	5.78
CBE	21.53	2.15	58.30	8.80	22.90	2.07	67.17	4.33
ILD	20.89	2.28	58.84	9.44	22.54	2.57	65.57	5.41
F value	321.255		115.242		292.065		317.966	
P value	< 0.001		< 0.001		< 0.001		< 0.001	

Table 6. Effect of rutile and antibiotics on hydrogen peroxide and delta amino levulinic acid

	H_2O_2 % (Increase with Rutile)		H_2O_2 % (Decrease with Doxy)		ALA % (Increase with Rutile)		ALA % (Decrease with Doxy)	
Normal	4.43	0.19	18.13	0.63	4.40	0.10	18.48	0.39
ASTH	22.90	1.51	63.27	4.96	23.43	1.57	66.30	3.57
CBE	23.43	1.74	64.28	7.33	22.76	2.20	67.63	3.52
ILD	23.30	1.47	60.35	7.93	22.63	1.63	67.24	3.42
F value	380.721		171.228		372.716		556.411	
P value	< 0.001		< 0.001		< 0.001		< 0.001	

Table 7. Effect of rutile and antibiotics on ATP synthase and cytochrome F 420

	ATP synthase % (Increase with Rutile)		ATP synthase % (Decrease with Doxy)		CYT F420 % (Increase with Rutile)		CYT F420 % (Decrease with Doxy)	
Normal	4.40	0.11	18.78	0.11	4.48	0.15	18.24	0.66
ASTH	22.94	1.94	56.18	4.15	22.39	1.75	64.24	8.55
CBE	23.32	1.74	55.67	4.16	22.78	2.23	62.58	8.62
ILD	23.33	1.35	56.83	3.27	21.95	1.56	53.17	7.20
F value	449.503		573.081		306.749		130.054	
P value	< 0.001		< 0.001		< 0.001		< 0.001	

The archaeal glycolytic hexokinase activity and archaeal extracellular ATP synthase activity were increased. The archaea can undergo magnetite and calcium carbonate mineralization and can exist as calcified nanoforms [18]. There was an increase in free RNA indicating self replicating RNA viroids and free DNA indicating generation of viroid complementary DNA strands by archaeal reverse transcriptase activity. The actinides modulate RNA folding and catalyse its ribozymal action. Digoxin can cut and paste the viroidal strands by modulating RNA splicing generating RNA viroidal diversity.

The viroids are evolutionarily escaped archaeal group I introns which have retrotransposition and self splicing qualities.[19] Archaeal pyruvate can produce histone deacetylase inhibition resulting in endogenous retroviral (HERV) reverse transcriptase and integrase expression. This can integrate the RNA viroidal complementary DNA into the noncoding region of eukaryotic non coding DNA using HERV integrase as has been described for borna and ebola viruses.[20] The noncoding DNA is lengthened by integrating RNA viroidal complementary DNA with the integration going on as a continuing event. The archaea genome can also get integrated into human genome using integrase as has been described for trypanosomes.[21] The integrated viroids and archaea can undergo vertical transmission and can exist as genomic parasites [20,21] This increases the length and alters the grammar of the noncoding region producing memes or memory of acquired characters [22] The viroidal complementary DNA can function as jumping genes producing a dynamic genome important in HLA gene expression. This modulation of HLA gene expression by viroidal complementary DNA can result in autoimmune diseases. The RNA viroids can regulate mrna function by RNA interference [19] The phenomena of RNA interference can modulate T cell and B cell function and euchromatin/heterochromatin expression. RNA viroidal mRNA interference plays a role in the pathogenesis of autoimmune diseases like interstitial lung disease, chronic bronchitis emphysema and bronchial asthma.

The presence of muramic acid, HMG CoA reductase and cholesterol oxidase activity inhibited by antibiotics indicates the presence of bacteria with mevalonate pathway. The bacterial with mevalonate pathway include streptococcus, staphylococcus, actinomycetes, listeria, coxiella and Borrelia.[23] The bacteria and archaea with mevalonate pathway and cholesterol catabolism had a evolutionarily advantage and constitutes the isoprenoidal clade organism with the archaea evolving into mevalonate pathway gram positive and gram negative organism through horizontal gene transfer of viroidal and virus genes [24] The isoprenoidal clade prokaryotes develop into other groups of prokaryotes via viroidal/virus as well as eukaryotic horizontal gene transfer producing bacterial speciation [25] The RNA viroids and its complementary DNA developed into cholesterol enveloped RNA and DNA viruses like herpes, retrovirus, influenza virus, borna virus, cytomegalo virus and Ebstein Barr virus by recombining with eukaryotic and human genes resulting in viral speciation. Bacterial and viral species are ill defined and fuzzy with all of them forming one common genetic pool with frequent horizontal gene transfer and recombination. Thus the multi and unicellular eukaryote with its genes serves the purpose of prokaryotic and viral speciation. The multicellular eukaryote developed so that their endosymbiotic archaeal colonies could survive and forage better. The multicellular eukaryotes are like bacterial biofilms. The archaea and bacteria with a mevalonate pathway uses the extracellular RNA viroids and DNA viroids for quorum sensing and in the generation of symbiotic biofilm like structures which develop into multicellular eukaryotes [26,27]. The endosymbiotic archaea and bacteria with mevalonate pathway still uses the RNA viroids and DNA viroids for the regulation of muticellular eukaryote. Pollution is induced by the primitive nanoarchaea and mevalonate pathway bacteria synthesised PAH and methane leading on to redox stress. Redox stress leads to sodium potassium ATPase inhibition, inward movement of plasma membrane cholesterol, defective SREBP sensing, increased cholesterol synthesis and nanoarchaeal/mevalonate pathway bacterial growth [28] Redox stress leads on to viroidal and archaeal multiplication. Redox stress can also lead to HERV reverse transcriptase and integrase expression. The noncoding DNA is formed of integrating RNA viroidal complementary DNA and archaea with the integration going on as a continuing event. The archaeal pox like dsDNA virus forms evolutionarily the nucleus. The integrated viroidal, archaeal and mevalonate pathway bacterial sequences can undergo vertical transmission and can exist as genomic parasites. The genomic integrated archaea, mevalonate pathway bacteria and viroids form a genomic reserve of bacteria and viruses which can recombine with human and eukaryotic genes producing bacterial and viral speciation. Bacteria and viruses have been related to the pathogenesis of interstitial lung disease, chronic bronchitis emphysema and bronchial asthma [29,30] The change in the length and grammar of the noncoding region produces eukaryotic speciation and individuality [31]. Changes in the length of noncoding region especially human endogenous retroviruses can lead onto autoimmune diseases [32] The integration of nanoarchaea, mevalonate pathway prokaryotes and viroids in to the eukaryotic and human genome produces a chimera which can multiply producing biofilm like multicellular structures having a mixed archaeal, viroidal, prokaryotic and eukaryotic characters which is a regression from the multicellular eukaryotic tissue This results in a new neuronal, metabolic, immune and tissue phenotype or microchimeras leading to human diseases like interstitial lung disease, chronic bronchitis emphysema and bronchial asthma. The microchimeras formed can lead to autoantigens and autoimmune diseases.

Archaea and RNA viroid can bind the TLR receptor induce NFKB producing immune activation and cytokine TNF alpha secretion. The archaeal DXP and mevalonate pathway metabolites can bind $\gamma\delta$ TCR and digoxin induced calcium signalling can activate NFKB producing chronic immune activation [2,33] The archaeal cholesterol aromatase generated PAH can produce immune activation. The archaea and viroid induced chronic immune activation and generation of superantigens can lead on to autoimmune disease. Immune activation has been related to the pathogenesis of interstitial lung disease, chronic bronchitis emphysema and bronchial asthma [29,30]. PAH has also been related to the pathogenesis of interstitial lung disease, chronic bronchitis emphysema and bronchial asthma.

The archaea and viroids can regulate the nervous system including the NMDA synaptic transmission [2]. NMDA can be activated by digoxin induced calcium oscillations, PAH and viroid induced RNA interference.[2] The cholesterol ring oxidase generated pyruvate can be converted by the GABA shunt pathway to glutamate. The archaeal cholesterol aromatase can generate serotonin [17]. Glutamatergic and serotoninergic transmission can lead to immune activation. Serotonin can produce bronchospasm leading onto bronchial asthma. The higher degree of integration of the archaea into the genome produces increased digoxin synthesis producing right hemispheric dominance and lesser degree producing left hemispheric dominance [2] Right hemispheric dominance can lead to autoimmune diseases like interstitial lung disease, chronic bronchitis emphysema and bronchial asthma [2]

Archaea, viroids and digoxin can induce the host AKT PI3K, AMPK, HIF alpha and NFKB producing the Warburg metabolic phenotype [34] The increased glycolytic hexokinase activity, decrease in blood ATP, leakage of cytochrome C, increase in serum pyruvate and decrease in acetyl CoA indicates the generation of the Warburg phenotype. There is induction of glycolysis, inhibition of PDH activity and mitochondrial dysfunction resulting in inefficient energetics. The lymphocytes depend on glycolysis for their energy needs. The increased glycolysis induced by the Warburg phenotype leads to immune activation. Lactic acid generated by increased glycolysis leads to immune stimulation. Cholesterol oxidase activity, increased glycolysis related NADPH oxidase activity, bacterial porphyrin induced redox stress and mitochondrial dysfunction generates free radicals important in the pathogenesis of interstitial lung disease, chronic bronchitis emphysema and bronchial asthma. The accumulated pyruvate enters the gaba shunt pathway and is converted to citrate which is acted upon by citrate lyase and converted to acetyl CoA, used for cholesterol synthesis [34] The pyruvate can be converted to glutamate and ammonia which is oxidised by archaea for energy needs. The increased cholesterol substrate leads to increased archaeal growth and digoxin synthesis leading to metabolic channelling to the mevalonate pathway. Hyperdigoxinemia is important in the pathogenesis of autoimmune diseases like interstitial lung disease, chronic bronchitis emphysema and bronchial asthma [2]. Digoxin can increase lymphocytic intracellular calcium which leads on to induction of NFKB and immune activation.[2] The archaeal cholesterol atabolism can deplete the lymphocytic cell membranes of cholesterol resulting in alteration of lymphocytic cell membrane microdomains related receptors producing immune activation. The archaeal bile acids can bind GPCR and modulate D2 regulating the conversion of T4 to T3. T3 activates uncoupling proteins reducing redox stress. Bile acids can also activate NRF ½ inducing NQO1, GST, HOI reducing redox stress. Bile acids can bind PXR inducing the bile acid shunt pathway of cholesterol detoxification. Bile acids can bind macrophage GPCR and VDR producing immunosuppression and inhibiting NFKB. This helps to modulate the archaea and viroid induced chronic immune activation. Bile acids are thus protective compounds and put a break on the archaea and viroid induced changes [35] Thus the actinide, viroid and mevalonate pathway bacteria induced metabolic, genetic, immune and neuronal transmission changes can lead onto interstitial lung disease, chronic bronchitis emphysema and bronchial asthma.

References

[1] Valiathan M.S., Somers, K., Kartha, C.C. (1993). *Endomyocardial Fibrosis.* Delhi: Oxford University Press.

[2] Kurup R., Kurup, P.A. (2009). *Hypothalamic digoxin, cerebral dominance and brain function in health and diseases*. New York: Nova Science Publishers.

[3] Hanold D., Randies, J.W. (1991). Coconut cadang-cadang disease and its viroid agent, *Plant Disease,* 75, 330-335.

[4] Edwin B.T., Mohankumaran, C. (2007). Kerala wilt disease phytoplasma: Phylogenetic analysis and identification of a vector, *Proutista moesta, Physiological and Molecular Plant Pathology,* 71(1-3), 41-47.

[5] Eckburg P.B., Lepp, P.W., Relman, D.A. (2003). Archaea and their potential role in human disease, *Infect Immun,* 71, 591-596.

[6] Adam Z. (2007). Actinides and Life's Origins, *Astrobiology,* 7, 6-10.

[7] Schoner W. (2002). Endogenous cardiac glycosides, a new class of steroid hormones, *Eur. J. Biochem.,* 269, 2440-2448.

[8] Davies P.C.W., Benner, S.A., Cleland, C.E., Lineweaver, C.H., McKay, C.P., Wolfe-Simon, F. (2009). Signatures of a Shadow Biosphere, *Astrobiology,* 10, 241-249.

[9] Richmond W. (1973). Preparation and properties of a cholesterol oxidase from nocardia species and its application to the enzymatic assay of total cholesterol in serum, *Clin. Chem.,* 19, 1350-1356.

[10] Snell E.D., Snell, C.T. (1961). *Colorimetric Methods of Analysis.* Vol 3A. New York: Van NoStrand.

[11] Glick D. (1971). *Methods of Biochemical Analysis.* Vol 5. New York: Interscience Publishers.

[12] Colowick, Kaplan, N.O. (1955). *Methods in Enzymology.* Vol 2. New York: Academic Press.

[13] Maarten A.H., Marie-Jose, M., Cornelia, G., van Helden–Meewsen, Fritz, E., Marten, P.H. (1995). Detection of muramic acid in human spleen, *Infection and Immunity,* 63(5), 1652 – 1657.

[14] Smit A., Mushegian, A. (2000). Biosynthesis of isoprenoids via mevalonate in Archaea: the lost pathway, *Genome Res.,* 10(10), 1468-84.

[15] Van der Geize R., Yam, K., Heuser, T., Wilbrink, M.H., Hara, H., Anderton, M.C. (2007). A gene cluster encoding cholesterol catabolism in a soil actinomycete provides insight into Mycobacterium tuberculosis survival in macrophages, *Proc. Natl. Acad. Sci. USA,* 104(6), 1947-52.

[16] Francis A.J. (1998). Biotransformation of uranium and other actinides in radioactive wastes, *Journal of Alloys and Compounds,* 271(273), 78-84.

[17] Probian C., Wülfing, A., Harder, J. (2003). Anaerobic mineralization of quaternary carbon atoms: Isolation of denitrifying bacteria on pivalic acid (2,2-Dimethylpropionic acid), *Applied and Environmental Microbiology,* 69(3), 1866-1870.

[18] Vainshtein M., Suzina, N., Kudryashova, E., Ariskina, E. (2002). New Magnet-Sensitive Structures in Bacterial and Archaeal Cells, *Biol. Cell,* 94(1), 29-35.

[19] Tsagris E.M., de Alba, A.E., Gozmanova, M., Kalantidis, K. (2008). Viroids, *Cell Microbiol,* 10, 2168.

[20] Horie M., Honda, T., Suzuki, Y., Kobayashi, Y., Daito, T., Oshida, T. (2010). Endogenous non-retroviral RNA virus elements in mammalian genomes, *Nature,* 463, 84-87.

[21] Hecht M., Nitz, N., Araujo, P., Sousa, A., Rosa, A., Gomes, D. (2010). Genes from Chagas parasite can transfer to humans and be passed on to children. Inheritance of DNA Transferred from American Trypanosomes to Human Hosts, *PLoS ONE,* 5, 2-10.

[22] Flam F. (1994). Hints of a language in junk DNA, *Science,* 266, 1320.

[23] Horbach S., Sahm, H., Welle, R. (1993). Isoprenoid biosynthesis in bacteria: two different pathways? *FEMS Microbiol Lett,* 111, 135–140.

[24] Gupta R.S. (1998). Protein phylogenetics and signature sequences: a reappraisal of evolutionary relationship among archaebacteria, eubacteria, and eukaryotes, *Microbiol. Mol. Biol. Rev.,* 62, 1435–1491.

[25] Hanage W., Fraser, C., Spratt, B. (2005). Fuzzy species among recombinogenic bacteria, *BMC Biology,* 3, 6-10.

[26] Webb J.S., Givskov, M., Kjelleberg, S. (2003). Bacterial biofilms: prokaryotic adventures in multicellularity, *Curr. Opin. Microbiol.*, 6(6), 578–85.

[27] Whitchurch C.B., Tolker-Nielsen, T., Ragas, P.C., Mattick, J.S. (2002). Extracellular DNA Required for Bacterial Biofilm Formation. *Science,* 295(5559), 1487.

[28] Chen Y., Cai, T., Wang, H., Li, Z., Loreaux, E., Lingrel, J.B. (2009). Regulation of intracellular cholesterol distribution by Na/K-ATPase, *J. Biol. Chem.*, 284(22), 14881-90.

[29] Fauci A.S. (1998). *Harrison's Principles of Internal Medicine*. New York: McGraw-Hill.

[30] Crystal R.G. (1996). *The Lung: Scientific Foundations*. Philadelphia: Lippincott-Raven.

[31] Poole A.M. (2006). Did group II intron proliferation in an endosymbiont-bearing archaeon create eukaryotes? *Biol. Direct,* 1, 36-40.

[32] Villarreal L.P. (2006). How viruses shape the tree of life, *Future Virology,* 1(5), 587-595.

[33] Eberl M., Hintz, M., Reichenberg, A., Kollas, A., Wiesner, J., Jomaa, H. (2010). Microbial isoprenoid biosynthesis and human γδ T cell activation, *FEBS Letters,* 544(1), 4-10.

[34] Wallace D.C. (2005). Mitochondria and Cancer: Warburg Addressed, *Cold Spring Harbor Symposia on Quantitative Biology,* 70, 363-374.

[35] Lefebvre P., Cariou, B., Lien, F., Kuipers, F., Staels, B. (2009). Role of Bile Acids and Bile Acid Receptors in Metabolic Regulation, *Physiol. Rev.,* 89(1), 147 - 191.

Chapter IX

A Cholesterol and Actinide Dependent Shadow Biosphere of Archaea and Viroids in Chronic Renal Failure

Abbreviations

CRF: Chronic renal failure

Introduction

Endomyocardial fibrosis (EMF) along with the root wilt disease of coconut is endemic to Kerala with its radioactive actinide beach sands. Actinides like rutile producing intracellular magnesium deficiency due to rutile-magnesium exchange sites in the cell membrane has been implicated in the etiology of EMF [1].

Endogenous digoxin, a steroidal glycoside which functions as a membrane sodium-potassium ATPase inhibitor has also been related to its etiology due to the intracellular magnesium deficiency it produces [2]. Organisms like phytoplasmas and viroids have also been demonstrated to play a role in the etiology of these diseases [3,4]. Endogenous digoxin has been related to the pathogenesis of chronic renal failure- chronic glomerulonephritis [2].

The possibility of endogenous digoxin synthesis by actinide based primitive organism like archaea with a mevalonate pathway and cholesterol catabolism was considered [5,6,7].

Davies has put forward the concept of a shadow biosphere of organisms with alternate biochemistry present in earth itself [8]. An actinide dependent shadow biosphere of archaea and viroids in the above mentioned disease states is described [6].

Materials and Methods

Informed consent of the subjects and the approval of the ethics committee were obtained for the study. The following groups were included in the study:- chronic renal failure- chronic glomerulonephritis. There were 10 patients in each group and each patient had an age and sex matched healthy control selected randomly from the general population. The blood samples were drawn in the fasting state before treatment was initiated. Plasma from fasting heparinised blood was used and the experimental protocol was as follows (I) Plasma+phosphate buffered saline, (II) same as I+cholesterol substrate, (III) same as II+rutile 0.1 mg/ml, (IV) same as II+ciprofloxacine and doxycycline each in a concentration of 1 mg/ml. Cholesterol substrate was prepared as described by Richmond [9]. Aliquots were withdrawn at zero time immediately after mixing and after incubation at 37°C for 1 hour. The following estimations were carried out:- Cytochrome F420, free RNA, free DNA, muramic acid, polycyclic aromatic hydrocarbon, hydrogen peroxide, serotonin, pyruvate, ammonia, glutamate, cytochrome C, hexokinase, ATP synthase, HMG CoA redutase, digoxin and bile acids [10,11,12,13]. Cytochrome F420 was estimated flourimetrically (excitation wavelength 420 nm and emission wavelength 520 nm). Polycyclic aromatic hydrocarbon was estimated by measuring hydrogen peroxide liberated by using glucose reagent. The statistical analysis was done by ANOVA.

Results

The parameters checked as indicated above were:- cytochrome F420, free RNA, free DNA, muramic acid, polycyclic aromatic hydrocarbon, hydrogen peroxide, serotonin, pyruvate, ammonia, glutamate, cytochrome C, hexokinase, ATP synthase, HMG CoA reductase, digoxin and bile acids. Plasma of control subjects showed increased levels of the above mentioned parameters with after incubation for 1 hour and addition of cholesterol substrate resulted in still further significant increase in these parameters. The plasma of patients showed similar results but the extent of increase was more. The addition of antibiotics to the control plasma caused a decrease in all the parameters while addition of rutile increased their levels.

The addition of antibiotics to the patient's plasma caused a decrease in all the parameters while addition of rutile increased their levels but the extent of change was more in patient's sera as compared to controls. The results are expressed in tables 1-7 as percentage change in the parameters after 1 hour incubation as compared to the values at zero time.

Discussion

There was increase in cytochrome F420 indicating archaeal growth in chronic renal failure- chronic glomerulonephritis. The archaea can synthesise and use cholesterol as a carbon and energy source [14,15]. The archaeal origin of the enzyme activities was indicated by antibiotic induced suppression. The study indicates the presence of actinide based archaea with an alternate actinide based enzymes or metalloenzymes in the system as indicated by rutile induced increase in enzyme activities [16]. There was also an increase in archaeal HMG CoA reductase activity indicating increased cholesterol synthesis by the archaeal mevalonate pathway. The archaeal beta hydroxyl steroid dehydrogenase activity indicating digoxin synthesis and archaeal cholesterol hydroxylase activity indicating bile acid synthesis were increased [7]. The archaeal cholesterol oxidase activity was increased resulting in generation of pyruvate and hydrogen peroxide [15]. The pyruvate gets converted to glutamate and ammonia by the GABA shunt pathway.

The archaeal aromatization of cholesterol generating PAH, serotonin and dopamine was also detected [17]. The archaeal glycolytic hexokinase activity and archaeal extracellular ATP synthase activity were increased. The archaea can undergo magnetite and calcium carbonate mineralization and can exist as calcified nanoforms [18]. There was an increase in free RNA indicating self replicating RNA viroids and free DNA indicating generation of viroid complementary DNA strands by archaeal reverse transcriptase activity. The actinides modulate RNA folding and catalyse its ribozymal action.

Digoxin can cut and paste the viroidal strands by modulating RNA splicing generating RNA viroidal diversity. The viroids are evolutionarily escaped archaeal group I introns which have retrotransposition and self splicing qualities [19]. Archaeal pyruvate can produce histone deacetylase inhibition resulting in endogenous retroviral (HERV) reverse transcriptase and integrase expression. This can integrate the RNA viroidal complementary DNA into the noncoding region of eukaryotic non coding DNA using HERV integrase as has been described for borna and ebola viruses [20]. The noncoding DNA is lengthened by integrating RNA viroidal complementary DNA with the integration going on as a continuing event. The archaea genome can also get integrated into human genome using integrase as has been described for trypanosomes [21]. The integrated viroids and archaea can undergo vertical transmission and can exist as genomic parasites [20,21]. This increases the length and alters the grammar of the noncoding region producing memes or memory of acquired characters [22]. The viroidal complementary DNA can function as jumping genes producing a dynamic genome important in HLA gene expression. This modulation of HLA gene expression by viroidal complementary DNA can result in immune activation. The RNA viroids can regulate mrna function by RNA interference [19]. The phenomena of RNA interference can modulate T cell and B cell function and euchromatin/heterochromatin expression. RNA viroidal mRNA interference plays a role in the pathogenesis of chronic renal failure- chronic glomerulonephritis due to immune activation.

The presence of muramic acid, HMG CoA reductase and cholesterol oxidase activity inhibited by antibiotics indicates the presence of bacteria with mevalonate pathway. The bacterial with mevalonate pathway include streptococcus, staphylococcus, actinomycetes, listeria, coxiella and borrelia [23]. The bacteria and archaea with mevalonate pathway and cholesterol catabolism had a evolutionarily advantage and constitutes the isoprenoidal clade organism with the archaea evolving into mevalonate pathway gram positive and gram negative organism through horizontal gene transfer of viroidal and virus genes [24]. The isoprenoidal clade prokaryotes develop into other groups of prokaryotes via viroidal/virus as well as eukaryotic horizontal gene transfer producing bacterial speciation [25]. The RNA viroids and its complementary DNA developed into cholesterol enveloped RNA and DNA viruses like herpes, retrovirus, influenza virus, borna virus, cytomegalo virus and Ebstein Barr virus by recombining with eukaryotic and human genes resulting in viral speciation.

Table 1. Effect of rutile and antibiotics on muramic acid and serotonin

	Muramic acid % change (Increase with Rutile)		Muramic acid % change (Decrease with Doxy+Cipro)		5 HT % (Increase without Doxy)		5 HT % (Decrease with Doxy)	
Group	Mean	± SD	Mean	± SD	Mean	± SD	Mean	± SD
Normal	4.41	0.15	18.63	0.12	4.34	0.15	18.24	0.37
CRF	23.41	1.55	66.36	4.31	23.49	1.19	64.63	6.58
F value	403.394		680.284		348.867		364.999	
P value	< 0.001		< 0.001		< 0.001		< 0.001	

Table 2. Effect of rutile and antibiotics on free DNA and RNA

	DNA % change (Increase with Rutile)		DNA % change (Decrease with Doxy)		RNA % change (Increase with Rutile)		RNA % change (Decrease with Doxy)	
Group	Mean	± SD	Mean	± SD	Mean	± SD	Mean	± SD
Normal	4.37	0.15	18.39	0.38	4.37	0.13	18.38	0.48
CRF	22.52	2.06	66.09	5.73	23.34	1.58	65.76	3.91
F value	337.577		356.621		427.828		654.453	
P value	< 0.001		< 0.001		< 0.001		< 0.001	

Table 3. Effect of rutile and antibiotics on HMG CoA reductase and PAH

	HMG CoA R % change (Increase with Rutile)		HMG CoA R % change (Decrease with Doxy)		PAH % change (Increase with Rutile)		PAH % change (Decrease with Doxy)	
Group	Mean	± SD	Mean	± SD	Mean	± SD	Mean	± SD
Normal	4.30	0.20	18.35	0.35	4.45	0.14	18.25	0.72
CRF	23.88	1.68	63.69	7.06	23.69	1.57	66.86	3.61
F value	319.332		199.553		391.318		257.996	
P value	< 0.001		< 0.001		< 0.001		< 0.001	

Table 4. Effect of rutile and antibiotics on digoxin and bile acids

	Digoxin (ng/ml) (Increase with Rutile)		Digoxin (ng/ml) (Decrease with Doxy+Cipro)		Bile acids % change (Increase with Rutile)		Bile acids % change (Decrease with Doxy)	
Group	Mean	± SD	Mean	± SD	Mean	± SD	Mean	± SD
Normal	0.11	0.00	0.[illegible]54	0.003	4.29	0.18	18.15	0.58
CRF	0.51	0.06	0.[illegible]92	0.035	23.29	1.41	62.44	7.64
F value	135.116		7[illegible]706		290.441		203.651	
P value	< 0.001		< [illegible].001		< 0.001		< 0.001	

Table 5. Effect on rutile and antibiotics on pyruvate and hexokinase

	Pyruvate % change (Increase with Rutile)		Pyruvate % change (Decrease with Doxy)		Hexokinase % change (Increase with Rutile)		Hexokinase % change (Decrease with Doxy)	
Group	Mean	± SD	Mean	± SD	Mean	± SD	Mean	± SD
Normal	4.34	0.21	1[illegible].43	0.82	4.21	0.16	18.56	0.76
CRF	21.00	2.02	6[illegible]03	7.33	22.95	1.49	65.72	4.58
F value	321.255		1[illegible].242		292.065		317.966	
P value	< 0.001		< [illegible].001		< 0.001		< 0.001	

Table 6. Effect of rutile and antibiotics on hydrogen peroxide and delta amino levulinic acid

	H_2O_2 % (Increase with Rutile)		H_2O_2 % (Decrease with Doxy)		ALA % (Increase with Rutile)		ALA % (Decrease with Doxy)	
Group	Mean	± SD	Mean	± SD	Mean	± SD	Mean	± SD
Normal	4.43	0.19	1[illegible].13	0.63	4.40	0.10	18.48	0.39
CRF	22.63	2.02	5[illegible].08	6.30	24.00	1.64	66.04	4.36
F value	380.721		1[illegible]1.228		372.716		556.411	
P value	< 0.001		< [illegible].001		< 0.001		< 0.001	

Table 7. Effect of rutile and antibiotics on ATP synthase and cytochrome F 420

	ATP synthase % (Increase with Rutile)		ATP synthase % (Decrease with Doxy)		CYT F420 % (Increase with Rutile)		CYT F420 % (Decrease with Doxy)	
Group	Mean	± SD	Mean	± SD	Mean	± SD	Mean	± SD
Normal	4.40	0.11	18.78	0.11	4.48	0.15	18.24	0.66
CRF	23.22	1.35	66.42	4.21	22.46	1.75	63.22	8.22
F value	449.503		673.081		306.749		130.054	
P value	< 0.001		< 0.001		< 0.001		< 0.001	

Bacterial and viral species are ill defined and fuzzy with all of them forming one common genetic pool with frequent horizontal gene transfer and recombination. Thus the multi and unicellular eukaryote with its genes serves the purpose of prokaryotic and viral speciation. The multicellular eukaryote developed so that their endosymbiotic archaeal colonies could survive and forage better. The multicellular eukaryotes are like bacterial biofilms. The archaea and bacteria with a mevalonate pathway uses the extracellular RNA viroids and DNA viroids for quorum sensing and in the generation of symbiotic biofilm like structures which develop into multicellular eukaryotes [26,27]. The endosymbiotic archaea and bacteria with mevalonate pathway still uses the RNA viroids and DNA viroids for the regulation of muticellular eukaryote. Pollution is induced by the primitive nanoarchaea and mevalonate pathway bacteria synthesised PAH and methane leading on to redox stress. Redox stress leads to sodium potassium ATPase inhibition, inward movement of plasma membrane cholesterol, defective SREBP sensing, increased cholesterol synthesis and nanoarchaeal/mevalonate pathway bacterial growth [28]. Redox stress leads on to viroidal and archaeal multiplication. Redox stress can also lead to HERV reverse transcriptase and integrase expression. The noncoding DNA is formed of integrating RNA viroidal complementary DNA and archaea with the integration going on as a continuing event. The archaeal pox like dsDNA virus forms evolutionarily the nucleus. The integrated viroidal, archaeal and mevalonate pathway bacterial sequences can undergo vertical transmission and can exist as genomic parasites. The genomic integrated archaea, mevalonate pathway bacteria and viroids form a genomic reserve of bacteria and viruses which can recombine with human and eukaryotic genes producing bacterial and viral speciation. Bacteria and viruses have been related to the pathogenesis of chronic renal failure- chronic glomerulonephritis [29,30]. The change in the length and grammar of the noncoding region produces eukaryotic speciation and individuality [31]. Changes in the length of noncoding region especially human endogenous retroviruses can lead onto immune activation and autoimmune diseases [32]. The integration of nanoarchaea, mevalonate pathway prokaryotes and viroids in to the eukaryotic and human genome produces a chimera which can multiply producing biofilm like multicellular structures having a mixed archaeal, viroidal, prokaryotic and eukaryotic characters which is a regression from the multicellular eukaryotic tissue This results in a new neuronal, metabolic, immune and tissue phenotype or microchimeras leading to human diseases like chronic renal failure- chronic glomerulonephritis. The microchimeras formed can lead to autoantigens and immune activation resulting in chronic renal failure- chronic glomerulonephritis.

Archaea and RNA viroid can bind the TLR receptor induce NFKB producing immune activation and cytokine TNF alpha secretion. The archaeal DXP and mevalonate pathway metabolites can bind γδ TCR and digoxin induced calcium signalling can activate NFKB producing chronic immune activation [2,33]. The archaeal cholesterol aromatase generated PAH can produce immune activation. The archaea and viroid induced chronic immune activation and generation of superantigens can lead on to immune activation and autoimmune disease. Immune activation has been related to the pathogenesis of chronic renal failure- chronic glomerulonephritis [29,30].

The archaea and viroids can regulate the nervous system including the NMDA synaptic transmission [2]. NMDA can be activated by digoxin induced calcium oscillations, PAH and viroid induced RNA interference [2]. The cholesterol ring oxidase generated pyruvate can be converted by the GABA shunt pathway to glutamate. The archaeal cholesterol aromatase can generate serotonin [17]. Glutamatergic and serotoninergic transmission can lead to immune activation. Immune activation mediated by neurotransmitters can contribute to chronic renal failure- chronic glomerulonephritis. The higher degree of integration of the archaea into the genome produces increased digoxin synthesis producing right hemispheric dominance and lesser degree producing left hemispheric dominance [2]. Right hemispheric dominance can lead to chronic renal failure- chronic glomerulonephritis [2].

Archaea, viroids and digoxin can induce the host AKT PI3K, AMPK, HIF alpha and NFKB producing the Warburg metabolic phenotype [34]. The increased glycolytic hexokinase activity, decrease in blood ATP, leakage of cytochrome C, increase in serum pyruvate and decrease in acetyl CoA indicates the generation of the Warburg phenotype. There is induction of glycolysis, inhibition of PDH activity and mitochondrial dysfunction resulting in inefficient energetics.

The lymphocytes depend on glycolysis for their energy needs. The increased glycolysis induced by the Warburg phenotype leads to immune activation. Lactic acid generated by increased glycolysis leads to immune stimulation. Immune activation consequent to the generation of the Warburg phenotype can lead to chronic renal failure- chronic glomerulonephritis. Cholesterol oxidase activity, increased glycolysis related NADPH oxidase activity, bacterial porphyrin induced redox stress and mitochondrial dysfunction generates free radicals important in the pathogenesis of chronic renal failure- chronic glomerulonephritis.

The accumulated pyruvate enters the GABA shunt pathway and is converted to citrate which is acted upon by citrate lyase and converted to acetyl CoA, used for cholesterol synthesis [34]. The pyruvate can be converted to glutamate and ammonia which is oxidised by archaea for energy needs. The increased cholesterol substrate leads to increased archaeal growth and digoxin synthesis leading to metabolic channelling to the mevalonate pathway. Hyperdigoxinemia is important in the pathogenesis of chronic renal failure- chronic glomerulonephritis [2]. Digoxin can increase lymphocytic intracellular calcium which leads on to induction of NFKB and immune activation [2].

The archaeal cholesterol catabolism can deplete the lymphocytic cell membranes of cholesterol resulting in alteration of lymphocytic cell membrane microdomains related receptors producing immune activation. Digoxin and membrane cholesterol depletion induced immune activation can contribute to chronic renal failure- chronic glomerulonephritis. The archaeal bile acids can bind GPCR and modulate D2 regulating the conversion of T4 to T3. T3 activates uncoupling proteins reducing redox stress. Bile acids can also activate NRF ½ inducing NQO1, GST, HOI reducing redox stress. Bile acids can bind PXR inducing the bile acid shunt pathway of cholesterol detoxification. Bile acids can bind macrophage GPCR and VDR producing immunosuppression and inhibiting NFKB.

This helps to modulate the archaea and viroid induced chronic immune activation. Bile acids are thus protective compounds and put a break on the archaea and viroid induced changes [35]. Thus the actinide, viroid and mevalonate pathway bacteria induced metabolic, genetic, immune and neuronal transmission changes can lead onto chronic renal failure- chronic glomerulonephritis.

References

[1] Valiathan M.S., Somers, K., Kartha, C.C. (1993). *Endomyocardial Fibrosis.* Delhi: Oxford University Press.

[2] Kurup R., Kurup, P.A. (2009). *Hypothalamic digoxin, cerebral dominance and brain function in health and diseases*. New York: Nova Science Publishers.

[3] Hanold D., Randies, J.W. (1991). Coconut cadang-cadang disease and its viroid agent, *Plant Disease,* 75, 330-335.

[4] Edwin B.T., Mohankumaran, C. (2007). Kerala wilt disease phytoplasma: Phylogenetic analysis and identification of a vector, *Proutista moesta, Physiological and Molecular Plant Pathology,* 71(1-3), 41-47.

[5] Eckburg P.B., Lepp, P.W., Relman, D.A. (2003). Archaea and their potential role in human disease, *Infect Immun,* 71, 591-596.

[6] Adam Z. (2007). Actinides and Life's Origins, *Astrobiology,* 7, 6-10.

[7] Schoner W. (2002). Endogenous cardiac glycosides, a new class of steroid hormones, *Eur. J. Biochem.,* 269, 2440-2448.

[8] Davies P.C.W., Benner, S.A., Cleland, C.E., Lineweaver, C.H., McKay, C.P., Wolfe-Simon, F. (2009). Signatures of a Shadow Biosphere, *Astrobiology,* 10, 241-249.

[9] Richmond W. (1973). Preparation and properties of a cholesterol oxidase from nocardia species and its application to the enzymatic assay of total cholesterol in serum, *Clin. Chem.,* 19, 1350-1356.

[10] Snell E.D., Snell, C.T. (1961). *Colorimetric Methods of Analysis.* Vol 3A. New York: Van NoStrand.

[11] Glick D. (1971). *Methods of Biochemical Analysis.* Vol 5. New York: Interscience Publishers.

[12] Colowick, Kaplan, N.O. (1955). *Methods in Enzymology.* Vol 2. New York: Academic Press.

[13] Maarten A.H., Marie Jose, M., Cornelis, G., van Helden Meeuwsen, Frits, E., Marten, P.H. (1995). Detection of muramic acid in human spleen, *Infection and Immunity,* 63(5), 1652 – 1657.

[14] Smit A., Mushegian, A. (2000). Biosynthesis of isoprenoids via mevalonate in Archaea: the lost pathway, *Genome Res.,* 10(10), 1468-84.

[15] Van der Geize R., Yam, K., Heuser, T., Wilbrink, M.H., Hara, H., Anderton, M.C. (2007). A gene cluster encoding cholesterol catabolism in a soil actinomycete provides insight into Mycobacterium tuberculosis survival in macrophages, *Proc. Natl. Acad. Sci. USA,* 104(6), 1947-52.

[16] Francis A.J. (1998). Biotransformation of uranium and other actinides in radioactive wastes, *Journal of Alloys and Compounds,* 271(273), 78-84.

[17] Probian C., Wülfing, A., Harder, J. (2003). Anaerobic mineralization of quaternary carbon atoms: Isolation of denitrifying bacteria on pivalic acid (2,2-Dimethylpropionic acid), *Applied and Environmental Microbiology,* 69(3), 1866-1870.

[18] Vainshtein M., Suzina, N., Kudryashova, E., Ariskina, E. (2002). New Magnet-Sensitive Structures in Bacterial and Archaeal Cells, *Biol. Cell,* 94(1), 29-35.

[19] Tsagris E.M., de Alba, A.E., Gozmanova, M., Kalantidis, K. (2008). Viroids, *Cell Microbiol.,* 10, 2168.

[20] Horie M., Honda, T., Suzuki, Y., Kobayashi, Y., Daito, T., Oshida, T. (2010). Endogenous non-retroviral RNA virus elements in mammalian genomes, *Nature,* 463, 84-87.

[21] Hecht M., Nitz, N., Araujo, P., Sousa, A., Rosa, A., Gomes, D. (2010). Genes from Chagas parasite can transfer to humans and be passed on to children. Inheritance of DNA Transferred from American Trypanosomes to Human Hosts, *PLoS ONE,* 5, 2-10.

[22] Flam F. (1994). Hints of a language in junk DNA, *Science,* 266, 1320.

[23] Horbach S., Sahm, H., Welle, R. (1993). Isoprenoid biosynthesis in bacteria: two different pathways? *FEMS Microbiol. Lett,* 111, 135–140.

[24] Gupta R.S. (1998). Protein phylogenetics and signature sequences: a reappraisal of evolutionary relationship among archaebacteria, eubacteria, and eukaryotes, *Microbiol. Mol. Biol. Rev.,* 62, 1435–1491.

[25] Hanage W., Fraser, C., Spratt, B. (2005). Fuzzy species among recombinogenic bacteria, *BMC Biology,* 3, 6-10.

[26] Webb J.S., Givskov, M., Kjelleberg, S. (2003). Bacterial biofilms: prokaryotic adventures in multicellularity, *Curr Opin Microbiol,* 6(6), 578–85.

[27] Whitchurch C.B., Tolker-Nielsen, T., Ragas, P.C., Mattick, J.S. (2002). Extracellular DNA Required for Bacterial Biofilm Formation. *Science,* 295(5559), 1487.

[28] Chen Y., Cai, T., Wang, H., Li, Z., Loreaux, E., Lingrel, J.B. (2009). Regulation of intracellular cholesterol distribution by Na/K-ATPase, *J. Biol. Chem.,* 284(22), 14881-90.

[29] Fauci A.S. (2009). Harrison's Principles of Internal Medicine. New York: McGraw-Hill.

[30] Brenner B.M. (2000). *The Kidney,* 5th edn. Philadelphia: Saunders.

[31] Poole A.M. (2006). Did group II intron proliferation in an endosymbiont-bearing archaeon create eukaryotes? *Biol. Direct,* 1, 36-40.

[32] Villarreal L.P. (2006). How viruses shape the tree of life, *Future Virology,* 1(5), 587-595.

[33] Eberl M., Hintz, M., Reichenberg, A., Kollas, A., Wiesner, J., Jomaa, H. (2010). Microbial isoprenoid biosynthesis and human γδ T cell activation, *FEBS Letters,* 544(1), 4-10.

[34] Wallace D.C. (2005). Mitochondria and Cancer: Warburg Addressed, *Cold Spring Harbor Symposia on Quantitative Biology,* 70, 363-374.

[35] Lefebvre P., Cariou, B., Lien, F., Kuipers, F., Staels, B. (2009). Role of Bile Acids and Bile Acid Receptors in Metabolic Regulation, *Physiol. Rev.,* 89(1), 147 - 191.

Chapter X

A Cholesterol and Actinide Dependent Shadow Biosphere of Archaea and Viroids in Metabolic Syndrome X

Abbreviations

DM: Type 2 diabetes mellitus
CAD: Coronary artery disease
CVA: Cerebrovascular thrombosis

Introduction

The human body synthesises an endogenous sodium-potassium ATPase inhibitor digoxin which plays a role in neuro-immuno-endocrine integration as well as in cardiovascular/ metabolic disorders. Endomyocardial fibrosis (EMF) along with the root wilt disease of coconut is endemic to Kerala with its radioactive actinide beach sands. Actinides like rutile producing intracellular magnesium deficiency due to rutile-magnesium exchange sites in the cell membrane has been implicated in the etiology of EMF [1]. Endogenous digoxin, a steroidal glycoside which functions as a membrane sodium-potassium ATPase inhibitor has also been related to its etiology due to the intracellular magnesium deficiency it produces [2]. Organisms like phytoplasmas and viroids have also been demonstrated to play a role in the etiology of these diseases [3,4]. Endogenous digoxin has een related to the pathogenesis of type 2 diabetes mellitus, coronary artery disease and cerebrovascular disease [2]. The possibility of endogenous digoxin synthesis by actinide based primitive organism like archaea with a mevalonate pathway and cholesterol catabolism was considered [5,6,7]. Davies has put forward the concept of a shadow biosphere of organisms with alternate biochemistry present in earth itself [8]. An actinide dependent shadow biosphere of archaea and viroids in the above mentioned disease states is described [6]. The intracellular endosymbionts archaea and their intron derived viroids constitute the third element regulating the human body.

Materials and Methods

Informed consent of the subjects and the approval of the ethics committee were obtained for the study. The following groups were included in the study:- type 2 diabetes mellitus, coronary artery disease- acute coronary syndrome and acute cerebrovascular thrombotic stroke. The coronary artery disease and cerebrovascular disease patients chosen for the study did not have type 2 diabetes mellitus as a risk factor. There were 10 patients in each group and each patient had an age and sex matched healthy control selected randomly from the general population. The blood samples were drawn in the fasting state before treatment was initiated. Plasma from fasting heparinised blood was used and the experimental protocol was as follows (I) Plasma+phosphate buffered saline, (II) same as I+cholesterol substrate, (III) same as II+rutile 0.1 mg/ml, (IV) same as II+ciprofloxacine and doxycycline each in a concentration of 1 mg/ml. Cholesterol substrate was prepared as described by Richmond [9]. Aliquots were withdrawn at zero time immediately after mixing and after incubation at 37°C for 1 hour. The following estimations were carried out:- Cytochrome F420, free RNA, free DNA, muramic acid, polycyclic aromatic hydrocarbon, hydrogen peroxide, serotonin, pyruvate, ammonia, glutamate, cytochrome C, hexokinase, ATP synthase, HMG CoA reductase, digoxin and bile acids [10,11,12,13]. Cytochrome F420 was estimated flourimetrically (excitation wavelength 420 nm and emission wavelength 520 nm). Polycyclic aromatic hydrocarbon was estimated by measuring hydrogen peroxide liberated by using glucose reagent. The statistical analysis was done by ANOVA.

Results

The parameters checked as indicated above were:- cytochrome F420, free RNA, free DNA, muramic acid, polycyclic aromatic hydrocarbon, hydrogen peroxide, serotonin, pyruvate, ammonia, glutamate, cytochrome C, hexokinase, ATP synthase, HMG CoA reductase, digoxin and bile acids. Plasma of control subjects showed increased levels of the above mentioned parameters with after incubation for 1 hour and addition of cholesterol substrate resulted in still further significant increase in these parameters. The plasma of patients showed similar results but the extent of increase was more. The addition of antibiotics to the control plasma caused a decrease in all the parameters while addition of rutile increased their levels. The addition of antibiotics to the patient's plasma caused a decrease in all the parameters while addition of rutile increased their levels but the extent of change was more in patient's sera as compared to controls. The results are expressed in tables 1-7 as percentage change in the parameters after 1 hour incubation as compared to the values at zero time.

Discussion

There was increase in cytochrome F420 indicating archaeal growth in type 2 diabetes mellitus, coronary artery disease and cerebrovascular disease. The archaea can synthesise and use cholesterol as a carbon and energy source [14,15]. The archaeal origin of the enzyme activities was indicated by antibiotic induced suppression. The study indicates the presence of actinide based archaea with an alternate actinide based enzymes or metalloenzymes in the system as indicated by rutile induced increase in enzyme activities [16]. There was also an increase in archaeal HMG CoA reductase activity indicating increased cholesterol synthesis by the archaeal mevalonate pathway. The archaeal beta hydroxyl steroid dehydrogenase activity indicating digoxin synthesis and archaeal cholesterol hydroxylase activity indicating bile acid synthesis were increased [7].

The archaeal cholesterol oxidase activity was increased resulting in generation of pyruvate and hydrogen peroxide [15]. The pyruvate gets converted to glutamate and ammonia by the GABA shunt pathway. The archaeal aromatization of cholesterol generating PAH, serotonin and dopamine was also detected [17]. The archaeal glycolytic hexokinase activity and archaeal extracellular ATP synthase activity were increased. The archaea can undergo magnetite and calcium carbonate mineralization and can exist as calcified nanoforms [18]. There was an increase in free RNA indicating self replicating RNA viroids and free DNA indicating generation of viroid complementary DNA strands by archaeal reverse transcriptase activity. The actinides modulate RNA folding and catalyse its ribozymal action. Digoxin can cut and paste the viroidal strands by modulating RNA splicing generating RNA viroidal diversity. The viroids are probably escaped archaeal group I introns which have retrotransposition and self splicing qualities [19]. The decrease in free self replicating RNA and DNA with the addition of antibiotics indicates that the RNA viroids are derived from archaeal introns. Archaeal pyruvate can produce histone deacetylase inhibition resulting in endogenous retroviral (HERV) reverse transcriptase and integrase expression. This can integrate the RNA viroidal complementary DNA into the noncoding region of eukaryotic non coding DNA using HERV integrase as has been described for borna and ebola viruses [20]. The noncoding DNA is lengthened by integrating RNA viroidal complementary DNA with the integration going on as a continuing event. The archaea genome can also get integrated into human genome using integrase as has been described for trypanosomes [21]. The integrated viroids and archaea can undergo vertical transmission and can exist as genomic parasites [20,21]. This increases the length and alters the grammar of the noncoding region producing memes or memory of acquired characters [22].

The viroidal complementary DNA can function as jumping genes producing a dynamic genome and changing DNA sequences. The type 2 diabetes mellitus has been related to altered genomic sequences mediated by HERV jumping genes. HERV particles have been related to the etiology of type 1 diabetes mellitus an autoimmune disease. The RNA viroids can regulate mRNA function by RNA interference [19]. The phenomena of RNA interference can modulate euchromatin/ heterochromatin expression. RNA viroidal mRNA interference plays a role in the pathogenesis of type 2 diabetes mellitus, coronary artery disease and cerebrovascular disease. Viroidal RNA mediated mRNA interference can modulate lipid metabolism triggering of dyslipidemias important in atherogenesis. The viroidal RNA modulation of T cell and B cell function by mRNA interference can lead to immune activation. Monocytic infiltration of the vascular wall is important in atherogenesis. Insulin resistance due to TNF alpha modulation of the insulin receptor can contribute to type 2 diabetes mellitus. The viroidal RNA mediated mRNA interference can also modulate insulin signalling and secretion leading onto type 2 diabetes mellitus.

Table 1. Effect of rutile and antibiotics on muramic acid and cytochrome F420

	Muramic acid % (Increase with Rutile)		Muramic acid % (Decrease with Doxy)		CYT F420 % (Increase with rutile)		CYT F420 % (Decrease with Doxy)	
Group	Mean	± SD	Mean	± SD	Mean	± SD	Mean	± SD
Normal	4.41	0.15	18.63	0.12	4.48	0.15	18.24	0.66
DM	24.10	1.61	65.78	4.43	22.59	1.86	57.05	8.45
CAD	23.34	1.75	66.80	3.43	22.76	2.26	60.49	6.86
CVA	22.94	1.99	68.30	2.52	21.01	2.29	62.37	8.01
F value	403.394		680.284		306.749		130.054	
P value	< 0.001		< 0.001		< 0.001		< 0.001	

Table 2. Effect of rutile and antibiotics on free DNA and RNA

	DNA % change (Increase with Rutile)		DNA % change (Decrease with Doxy)		RNA % change (Increase with Rutile)		RNA % change (Decrease with Doxy)	
Group	Mean	± SD	Mean	± SD	Mean	± SD	Mean	± SD
Normal	4.37	0.15	18.39	0.38	4.37	0.13	18.38	0.48
DM	23.01	1.67	65.35	3.56	23.33	1.86	66.46	3.65
CAD	23.12	1.71	65.12	5.58	24.01	1.17	66.66	3.84
CVA	22.51	1.85	63.56	5.29	22.95	1.90	66.39	3.83
F value	337.577		356.621		427.828		654.453	
P value	< 0.001		< 0.001		< 0.001		< 0.001	

Table 3. Effect of rutile and antibiotics on HMG CoA reductase and PAH

	HMG CoA R % change (Increase with Rutile)		HMG CoA R % change (Decrease with Doxy)		PAH % change (Increase with Rutile)		PAH % change (Decrease with Doxy)	
Group	Mean	± SD	Mean	± SD	Mean	± SD	Mean	± SD
Normal	4.30	0.20	18.35	0.35	4.45	0.14	18.25	0.72
DM	23.06	1.65	62.25	6.24	23.40	1.55	65.77	5.27
CAD	23.63	1.58	61.19	7.03	22.22	2.33	61.73	6.33
CVA	22.51	2.47	60.77	5.89	23.87	1.64	66.01	5.78
F value	319.332		199.553		391.318		257.996	
P value	< 0.001		< 0.001		< 0.001		< 0.001	

Table 4. Effect of rutile and antibiotics on digoxin and bile acids

	Digoxin (ng/ml) (Increase with Rutile)		Digoxin (ng/ml) (Decrease with Doxy+Cipro)		Bile acids % change (Increase with Rutile)		Bile acids % change (Decrease with Doxy)	
Group	Mean	± SD	Mean	± SD	Mean	± SD	Mean	± SD
Normal	0.11	0.00	0.054	0.003	4.29	0.18	18.15	0.58
DM	0.47	0.04	0.202	0.025	22.87	2.58	64.51	5.93
CAD	0.49	0.07	0.202	0.021	22.22	2.44	63.47	6.98
CVA	0.51	0.07	0.195	0.023	22.33	2.18	62.20	6.33
F value	135.116		7[illegible].706		290.441		203.651	
P value	< 0.001		< 0.001		< 0.001		< 0.001	

Table 5. Effect of rutile and antibiotics on pyruvate and hexokinase

	Pyruvate % change (Increase with Rutile)		Pyruvate % change (Decrease with Doxy)		Hexokinase % change (Increase with Rutile)		Hexokinase % change (Decrease with Doxy)	
Group	Mean	± SD	Mean	± SD	Mean	± SD	Mean	± SD
Normal	4.34	0.21	13.43	0.82	4.21	0.16	18.56	0.76
DM	20.67	1.38	53.75	8.12	23.23	1.88	65.11	5.14
CAD	20.16	1.07	57.08	9.83	21.88	2.11	65.02	4.40
CVA	20.60	1.81	53.97	7.03	21.98	2.12	65.78	6.08
F value	321.255		[illegible]5.242		292.065		317.966	
P value	< 0.001		< 0.001		< 0.001		< 0.001	

Table 6. Effect of rutile and antibiotics on ATP synthase and hydrogen peroxide

	ATP synthase % (Increase with Rutile)		ATP synthase % (Decrease with Doxy)		H_2O_2 % (Increase with Rutile)		H_2O_2 % (Decrease with Doxy)	
Group	Mean	± SD	Mean	± SD	Mean	± SD	Mean	± SD
Normal	4.40	0.11	18.78	0.11	4.43	0.19	18.13	0.63
DM	23.72	1.73	56.25	3.69	23.27	1.53	58.91	6.09
CAD	23.78	1.20	56.90	4.10	23.24	1.85	57.08	7.42
CVA	23.47	1.60	56.27	3.88	23.32	1.60	61.45	7.01
F value	449.503		573.081		380.721		171.228	
P value	< 0.001		< 0.001		< 0.001		< 0.001	

Table 7. Effect of rutile and antibiotics on serotonin and delta amino levulinic acid

	5 HT % (Increase with Rutile)		5 HT % (Decrease with Doxy)		PORPHYRIN % (Increase with Rutile)		PORPHYRIN % (Decrease with Doxy)	
Group	Mean	± SD	Mean	± SD	Mean	± SD	Mean	± SD
Normal	4.34	0.15	18.24	0.37	4.40	0.10	18.48	0.39
DM	22.73	2.46	65.87	4.35	22.87	1.84	66.31	3.68
CAD	22.42	1.99	61.14	3.47	23.81	1.90	66.95	3.67
CVA	23.35	1.83	63.73	6.52	23.73	1.70	66.44	3.92
F value	348.867		364.999		372.716		556.411	
P value	< 0.001		< 0.001		< 0.001		< 0.001	

The presence of muramic acid, HMG CoA reductase and cholesterol oxidase activity inhibited by antibiotics indicates the presence of bacteria with mevalonate pathway. The bacterial with mevalonate pathway include streptococcus, staphylococcus, actinomycetes, listeria, coxiella and borrelia[23]. The bacteria and archaea with mevalonate pathway and cholesterol catabolism had a evolutionarily advantage and constitutes the isoprenoidal clade organism with the archaea evolving into mevalonate pathway gram positive and gram negative organism through horizontal gene transfer of viroidal and virus genes [24]. The isoprenoidal clade prokaryotes develop into other groups of prokaryotes via viroidal/virus as well as eukaryotic horizontal gene transfer producing bacterial speciation [25]. The RNA viroids and its complementary DNA developed into cholesterol enveloped RNA and DNA viruses like herpes, retrovirus, influenza virus, borna virus, cytomegalo virus and Ebstein Barr virus by recombining with eukaryotic and human genes resulting in viral speciation. Bacterial and viral species are ill defined and fuzzy with all of them forming one common genetic pool with frequent horizontal gene transfer and recombination. Thus the multi and unicellular eukaryote with its genes serves the purpose of prokaryotic and viral speciation. The multicellular eukaryote developed so that their endosymbiotic archaeal colonies could survive and forage better. The multicellular eukaryotes are like bacterial biofilms. The archaea and bacteria with a mevalonate pathway uses the extracellular RNA viroids and DNA viroids for quorum sensing and in the generation of symbiotic biofilm like structures which develop into multicellular eukaryotes [26,27]. The endosymbiotic archaea and bacteria with mevalonate pathway still uses the RNA viroids and DNA viroids for the regulation of muticellular eukaryote. Pollution is induced by the primitive nanoarchaea and mevalonate pathway bacteria synthesised PAH and methane leading on to redox stress. Redox stress leads to sodium potassium ATPase inhibition, inward movement of plasma membrane cholesterol, defective SREBP sensing, increased cholesterol synthesis and nanoarchaeal/mevalonate pathway bacterial growth [28]. Redox stress leads on to viroidal and archaeal multiplication. Redox stress can also lead to HERV reverse transcriptase and integrase expression. The noncoding DNA is formed of integrating RNA viroidal complementary DNA and archaea with the integration going on as a continuing event. The archaeal pox like dsDNA virus forms evolutionarily the nucleus. The integrated viroidal, archaeal and mevalonate pathway bacterial sequences can undergo vertical transmission and can exist as genomic parasites. The genomic integrated archaea, mevalonate pathway bacteria and viroids form a genomic reserve of bacteria and viruses which can recombine with human and eukaryotic genes producing bacterial and viral speciation. Bacteria and viruses have been related to the pathogenesis of atherogenesis and type 2 diabetes mellitus. Archaea, chlamydia, rickettsia, mycoplasma, cytomegalovirus and herpes virus has been related to the etiology of atherosclerosis [29]. Type 2 diabetes mellitus is related to an acute phase response and TNF alpha mediated insulin resistance. Persistent bacterial and viroid symbiosis can lead to immune activation and TNF alpha mediated insulin resistance [30]. The change in the length and grammar of the noncoding region produces eukaryotic speciation and individuality [31]. Changes in the length of noncoding region especially human endogenous retroviruses can lead onto type 2 diabetes mellitus, coronary artery disease and cerebrovascular disease[32]. HERV sequences functioning as jumping genes can produce alteration in DNA sequences contributing to type 2 diabetes mellitus [32]. The viroid complementary DNA integrated into the genome can also function as jumping genes producing new genomic sequences leading onto type 2 diabetes mellitus. The integration of nanoarchaea, mevalonate pathway prokaryotes and viroids in to the eukaryotic and human genome produces a chimera which can multiply producing biofilm like multicellular structures having a mixed archaeal, viroidal, prokaryotic and eukaryotic characters which is a regression from the multicellular eukaryotic tissue. This results in a new metabolic and immune phenotype leading to human diseases like type 2 diabetes mellitus, coronary artery disease and cerebrovascular disease. The microchimeras formed can lead to atherogenesis and type 2 diabetes mellitus.

Archaea and RNA viroid can bind the TLR receptor induce NFKB producing immune activation and cytokine TNF alpha secretion. The archaeal DXP and mevalonate pathway metabolites can bind $\gamma\delta$ TCR and digoxin induced calcium signaling can activate NFKB producing chronic immune activation [2,33]. The archaea and viroid induced chronic immune activation and generation of superantigens. Immune activation results in induction of NADPH oxidase which generates hydrogen peroxide. Cholesterol oxidase activity also generates hydrogen peroxide. Hydrogen peroxide can increase protein tyrosine kinase activity and suppress protein phosphatase activity increasing insulin receptor function. Immune activated NOX and bacterial cholesterol oxidase can thus regulate insulin receptor function. Immune activation can also produce insulin resistance. TNF alpha produced by chronic immune activation can modulate the insulin receptor producing insulin resistance [30]. Chronic immune activation and cholesterol oxidase generated hydrogen peroxide can induce neutral sphingomyelinase generating ceramide producing insulin resistance [34]. Immune activation and NFKB induction can suppress the nuclear receptors LXR, PXR and FXR. LXR suppression by NFKB stimulates HMG CoA reductase activity and suppresses cholesterol 7 alpha hydroxylase activity [35]. This stimulates cholesterol synthesis and inhibits its degradation via the bile acid pathway. PXR suppression by NFKB prevents cholesterol detoxification via the bile acid shunt pathway [36]. Thus LXR and PXR suppression by NFKB produces acute cholesterol toxicity. This NFKB induced suppression of LXR and PXR can contribute to increased lipid and cholesterol synthesis contributing to obesity. FXR suppression can also lead to insulin resistance, dyslipidemias and increased connective tissue MPS deposition in vessel wall and atherogenesis. The cholesterol toxicity can lead to lipoprotein and cholesterol uptake by monocytes in the vessel wall producing atherogenesis. The archaea and viroid induced chronic immune activation can lead to monocyte infiltration of the vessel wall. This sets the stage for the atherogenetic process [29]. The increased cholesterol synthesis is important in stimulating archaeal growth which uses cholesterol as a carbon and energy source. Archaea, viroids and digoxin can induce the host AKT PI3K, AMPK, HIF alpha and NFKB producing the Warburg metabolic phenotype [37]. The increased glycolytic hexokinase activity, decrease in blood ATP, leakage of cytochrome C, increase in serum pyruvate and decrease in acetyl CoA indicates the generation of the Warburg phenotype. There is induction of glycolysis, inhibition of PDH activity and mitochondrial dysfunction resulting in inefficient energetics. Inefficient energetics owing to the Warburg's phenotype can contribute to metabolic syndrome X. The accumulated pyruvate enters the gaba shunt pathway and is converted to citrate which is acted upon by citrate lyase and converted to acetyl CoA, used for cholesterol synthesis [37]. The pyruvate can be converted to glutamate and ammonia which is oxidised by archaea for energy needs. Ammonia can stimulate membrane sodium-potassium ATPase increasing ATP utilisation, produce mitochondrial transmembrane potential changes and produce mitochondrial dysfunction important in type 2 diabetes mellitus. The increased cholesterol substrate also leads to increased archaeal growth and digoxin synthesis due to metabolic channeling to the mevalonate pathway. Digoxin can produce sodium-potassium ATPase inhibition and inward movement of plasma membrane cholesterol. This produces defective SREBP sensing, increased HMG CoA reductase activity and cholesterol synthesis [28]. The digoxin induced inward movement of plasma membrane cholesterol can alter membrane cholesterol/sphingomyelin ratio producing modified lipid microdomains [38]. The digoxin induced lipid microdomain modulation can regulate the GPCR couple adrenaline, noradrenaline, glucagon and neuropeptide receptors as well as protein tyrosine kinase linked insulin receptor. The digoxin mediated inhibition of nuclear membrane sodium-potassium ATPase can modulate nuclear membrane lipid microdomains and steroidal/thyroxine DNA receptor function. Thus endogenous digoxin can modulate all the endocrine receptors by regulating lipid microdomains. Hyperdigoxinemia is important in the pathogenesis of atherogenesis and metabolic syndrome X. Digoxin induced sodium-potassium ATPase inhibition results in an ATP sparing effect [39]. Eighty percent of the ATP generated is used to run the sodium-potassium ATPase pump. The digoxin inhibition of the sodium-potassium ATPase spares this ATP which is then used for lipid synthesis. Thus endogenous digoxin and the shadow biosphere generated Warburg phenotype can produce increased lipid synthesis and obesity important in metabolic syndrome X. Fat fuels insulin resistance by binding to the toll receptor and producing immune activation and immune infiltration of the adipose tissue. Digoxin can also increase lymphocytic intracellular calcium which leads on to induction of NFKB and immune activation [2]. The archaeal cholesterol catabolism can deplete the lymphocytic cell membranes of cholesterol resulting in alteration of lymphocytic cell membrane microdomains related receptors producing immune activation. The archaeal bile acids are steroidal hormones [40]. The archaeal bile acids can bind GPCR and modulate D2 regulating the conversion of T4 to T3. T3 activates uncoupling proteins reducing redox stress. Bile acids can also activate NRF ½ inducing NQO1, GST, HOI reducing redox stress. Bile acids can bind FXR regulating insulin receptor sensitivity and bind PXR inducing the bile acid shunt pathway of cholesterol detoxification. Bile acids can bind macrophage GPCR and VDR producing immunosuppression and inhibiting NFKB. This helps to modulate the archaea and viroid induced chronic immune activation. Thus the archaeal bile acids have a role opposite to digoxin and help to increase insulin sensitivity. The archaea and viroids can regulate the nervous system including the NMDA synaptic transmission [2]. NMDA can be activated by digoxin induced calcium oscillations, PAH and viroid induced RNA interference [2]. The cholesterol ring oxidase generated pyruvate can be converted by the GABA shunt pathway to glutamate. Glutamatergic transmission can lead to immune activation, atherogenesis and increased insulin signalling/release. The archaeal cholesterol aromatase can generate PAH [17]. The PAH can also lead to insulin resistance and atherogenesis. Particulate pollution has been related to metabolic syndrome X, type 2 diabetes mellitus and vascular thrombosis. The aromatase generated serotonin and dopamine can lead onto mood disorders common in type 2 diabetes mellitus. The increased cholesterol catabolism generated glutamate can lead onto NMDA excitotoxicity and archaea/viroids can lead onto reactive amyloid deposition important in the pathogenesis of alzheimer's disease common in type 2 diabetes mellitus. The Warburg phenotype can lead onto oncogenesis common in type 2 diabetes mellitus. The higher degree of integration of the archaea into the genome produces increased digoxin synthesis producing right hemispheric dominance and lesser degree producing left hemispheric dominance [2]. Right hemispheric dominance can lead to type 2 diabetes mellitus and vascular thrombosis. Thus the actinide, viroid and mevalonate pathway bacteria induced metabolic, genetic, immune and neuronal transmission changes can lead onto metabolic syndrome X and atherogenesis.

An actinide dependent shadow biosphere of archaea and viroids is described in type 2 diabetes mellitus, coronary artery disease- acute coronary syndrome and acute cerebrovascular thrombosis contributing to their pathogenesis. The archaea secreted digoxin serves as a messenger regulating metabolic and endocrine systems.

References

[1] Valiathan M.S., Somers, K., Kartha, C.C. (1993). *Endomyocardial Fibrosis.* Delhi: Oxford University Press.

[2] Kurup R., Kurup, P.A. (2009). *Hypothalamic digoxin, cerebral dominance and brain function in health and diseases.* New York: Nova Science Publishers.

[3] Hanold D., Randies, J.W. (1991). Coconut cadang-cadang disease and its viroid agent, *Plant Disease,* 75, 330-335.

[4] Edwin B.T., Mohankumaran, C. (2007). Kerala wilt disease phytoplasma: Phylogenetic analysis and identification of a vector, *Proutista moesta, Physiological and Molecular Plant Pathology,* 71(1-3), 41-47.

[5] Eckburg P.B., Lepp, P.W., Relman, D.A. (2003). Archaea and their potential role in human disease, *Infect. Immun.,* 71, 591-596.

[6] Adam Z. (2007). Actinides and Life's Origins, *Astrobiology,* 7, 6-10.

[7] Schoner W. (2002). Endogenous cardiac glycosides, a new class of steroid hormones, *Eur. J. Biochem.,* 269, 2440-2448.

[8] Davies P.C.W., Benner, S.A., Cleland, C.E., Lineweaver, C.H., McKay, C.P., Wolfe-Simon, F. (2009). Signatures of a Shadow Biosphere, *Astrobiology,* 10, 241-249.

[9] Richmond W. (1973). Preparation and properties of a cholesterol oxidase from nocardia species and its application to the enzymatic assay of total cholesterol in serum, *Clin. Chem.,* 19, 1350-1356.

[10] Snell E.D., Snell, C.T. (1961). *Colorimetric Methods of Analysis.* Vol 3A. New York: Van NoStrand.

[11] Glick D. (1971). *Methods of Biochemical Analysis.* Vol 5. New York: Interscience Publishers.

[12] Colowick, Kaplan, N.O. (1955). *Methods in Enzymology.* Vol 2. New York: Academic Press.

[13] Maarten A.H., Marie-Jose, M., Cornelia, G., van Helden–Meewsen, Fritz, E., Marten, P.H. (1995). Detection of muramic acid in human spleen, *Infection and Immunity,* 63(5), 1652 – 1657.

[14] Smit A., Mushegian, A. (2000). Biosynthesis of isoprenoids via mevalonate in Archaea: the lost pathway, *Genome Res.,* 10(10), 1468-84.

[15] Van der Geize R., Yam, K., Heuser, T., Wilbrink, M.H., Hara, H., Anderton, M.C. (2007). A gene cluster encoding cholesterol catabolism in a soil actinomycete provides insight into Mycobacterium tuberculosis survival in macrophages, *Proc. Natl. Acad. Sci. USA,* 104(6), 1947-52.

[16] Francis A.J. (1998). Biotransformation of uranium and other actinides in radioactive wastes, *Journal of Alloys and Compounds,* 271(273), 78-84.

[17] Probian C., Wülfing, A., Harder, J. (2003). Anaerobic mineralization of quaternary carbon atoms: Isolation of denitrifying bacteria on pivalic acid (2,2-Dimethylpropionic acid), *Applied and Environmental Microbiology,* 69(3), 1866-1870.

[18] Vainshtein M., Suzina, N., Kudryashova, E., Ariskina, E. (2002). New Magnet-Sensitive Structures in Bacterial and Archaeal Cells, *Biol. Cell,* 94(1), 29-35.

[19] Tsagris E.M., de Alba, A.E., Gozmanova, M., Kalantidis, K. (2008). Viroids, *Cell Microbiol.,* 10, 2168.

[20] Horie M., Honda, T., Suzuki, Y., Kobayashi, Y., Daito, T., Oshida, T. (2010). Endogenous non-retroviral RNA virus elements in mammalian genomes, *Nature,* 463, 84-87.

[21] Hecht M., Nitz, N., Araujo, P., Sousa, A., Rosa, A., Gomes, D. (2010). Genes from Chagas parasite can transfer to humans and be passed on to children. Inheritance of DNA Transferred from American Trypanosomes to Human Hosts, *PLoS ONE,* 5, 2-10.

[22] Flam F. (1994). Hints of a language in junk DNA, *Science,* 266, 1320.

[23] Horbach S., Sahm, H., Welle, R. (1993). Isoprenoid biosynthesis in bacteria: two different pathways? *FEMS Microbiol. Lett,* 111, 135–140.

[24] Gupta R.S. (1998). Protein phylogenetics and signature sequences: a reappraisal of evolutionary relationship among archaebacteria, eubacteria, and eukaryotes, *Microbiol Mol Biol Rev,* 62, 1435–1491.

[25] Hanage W., Fraser, C., Spratt, B. (2005). Fuzzy species among recombinogenic bacteria, *BMC Biology,* 3, 6-10.

[26] Webb J.S., Givskov, M., Kjelleberg, S. (2003). Bacterial biofilms: prokaryotic adventures in multicellularity, *Curr. Opin. Microbiol.,* 6(6), 578–85.

[27] Whitchurch C.B., Tolker-Nielsen, T., Ragas, P.C., Mattick, J.S. (2002). Extracellular DNA Required for Bacterial Biofilm Formation. *Science,* 295(5559), 1487.

[28] Chen Y., Cai, T., Wang, H., Li, Z., Loreaux, E., Lingrel, J.B. (2009). Regulation of intracellular cholesterol distribution by Na/K-ATPase, *J. Biol. Chem.,* 284(22), 14881-90.

[29] Khovidhunkit W., Kim, M.S., Memon, R.A., Shigenaga, J.K., Moser, A.H., Feingold, K.R. (2004). Thematic review series: The Pathogenesis of Atherosclerosis. Effects of infection and inflammation on lipid and lipoprotein metabolism mechanisms and consequences to the host. *J. Lip Res.,* 45(7), 1169 - 1196.

[30] Cani P.D., Amar, J., Iglesias, M.A., Poggi, M., Knauf, C., Bastelica, D. (2007). Metabolic endotoxemia initiates obesity and insulin resistance. *Diabetes,* 56, 1761–1772.

[31] Poole A.M. (2006). Did group II intron proliferation in an endosymbiont-bearing archaeon create eukaryotes? *Biol. Direct,* 1, 36-40.

[32] Villarreal L.P. (2006). How viruses shape the tree of life, *Future Virology,* 1(5), 587-595.

[33] Eberl M., Hintz, M., Reichenberg, A., Kollas, A., Wiesner, J., Jomaa, H. (2010). Microbial isoprenoid biosynthesis and human γδ T cell activation, *FEBS Letters,* 544(1), 4-10.

[34] Memon R.A., Holleran, W.M., Moser, A.H., Seki, T., Uchida, Y., Fuller, J. (1998). Endotoxin and Cytokines Increase Hepatic Sphingolipid Biosynthesis and Produce

Lipoproteins Enriched in Ceramides and Sphingomyelin. *Arterioscler. Thromb. Vasc. Biol.,* 18(8), 1257-1265.

[35] Carayol N., Chen, J., Yang, F., Jin, T., Jin, L., States, D. (2006). A dominant function of IKK/NF-kB signaling in global LPS-induced gene expression. *J. Biol. Chem.,* 10, 1074.

[36] Kliewer S.A. (2005). Cholesterol detoxification by the nuclear pregnane X receptor. *Proc. Natl. Acad. Sci. USA,* 102(8), 2675-6.

[37] Wallace D.C. (2005). Mitochondria and Cancer: Warburg Addressed, *Cold Spring Harbor Symposia on Quantitative Biology,* 70, 363-374.

[38] Paila Y.D., Tiwari, S., Chattopadhyay, A. (2009). Are specific nonannular cholesterol binding sites present in G-protein coupled receptors? *Biochim. Biophys. Acta,* 1788(2), 295-302.

[39] Ebensperger G., Ebensperger, R., Herrera, E.A., Riquelme, R.A., Sanhueza, E.M., Lesage, F. (2005). Fetal brain hypometabolism during prolonged hypoxaemia in the llama. *J. Physiol.,* 567(3), 963 – 975.

[40] Lefebvre P., Cariou, B., Lien, F., Kuipers, F., Staels, B. (2009). Role of Bile Acids and Bile Acid Receptors in Metabolic Regulation, *Physiol. Rev.,* 89(1), 147 - 191.

Chapter XI

A Cholesterol and Actinide Dependent Shadow Biosphere of Archaea and Viroids in Autoimmune Diseases

Abbreviations

MS: Multiple sclerosis
SLE: Systemic lupus erythematosis
RA: Rheumatoid arthritis

Introduction

Endomyocardial fibrosis (EMF) along with the root wilt disease of coconut is endemic to Kerala with its radioactive actinide beach sands. Actinides like rutile producing intracellular magnesium deficiency due to rutile-magnesium exchange sites in the cell membrane has been implicated in the etiology of EMF[1] Endogenous digoxin, a steroidal glycoside which functions as a membrane sodium-potassium ATPase inhibitor has also been related to its etiology due to the intracellular magnesium deficiency it produces [2]. Organisms like phytoplasmas and viroids have also been demonstrated to play a role in the etiology of these diseases [3,4]. Endogenous digoxin has been related to the pathogenesis of multiple sclerosis and other autoimmune diseases like systemic lupus erythematosis and rheumatoid arthritis [2].

The possibility of endogenous digoxin synthesis by actinide based primitive organism like archaea with a mevalonate pathway and cholesterol catabolism was considered [5,6,7]. Davies has put forward the concept of a shadow biosphere of organisms with alternate biochemistry present in earth itself [8]. An actinide dependent shadow biosphere of archaea and viroids in the above mentioned disease states is described [6].

Materials and Methods

Informed consent of the subjects and the approval of the ethics committee were obtained for the study. The following groups were included in the study:- multiple sclerosis, systemic lupus erythematosis and rheumatoid arthritis. There were 10 patients in each group and each patient had an age and sex matched healthy control selected randomly from the general population.

The blood samples were drawn in the fasting state before treatment was initiated. Plasma from fasting heparinised blood was used and the experimental protocol was as follows (I) Plasma+phosphate buffered saline, (II) same as I+cholesterol substrate, (III) same as II+rutile 0.1 mg/ml, (IV) same as II+ciprofloxacine and doxycycline each in a concentration of 1 mg/ml. Cholesterol substrate was prepared as described by Richmond [9]. Aliquots were withdrawn at zero time immediately after mixing and after incubation at 37°C for 1 hour.

The following estimations were carried out:- Cytochrome F420, free RNA, free DNA, muramic acid, polycyclic aromatic hydrocarbon, hydrogen peroxide, serotonin, pyruvate, ammonia, glutamate, cytochrome C, hexokinase, ATP synthase, HMG CoA redutase, digoxin and bile acids [10,11,12,13]. Cytochrome F420 was estimated flourimetrically (excitation wavelength 420 nm and emission wavelength 520 nm). Polycyclic aromatic hydrocarbon was estimated by measuring hydrogen peroxide liberated by using glucose reagent. The statistical analysis was done by ANOVA.

Results

The parameters checked as indicated above were:- cytochrome F420, free RNA, free DNA, muramic acid, polycyclic aromatic hydrocarbon, hydrogen peroxide, serotonin, pyruvate, ammonia, glutamate, cytochrome C, hexokinase, ATP synthase, HMG CoA reductase, digoxin and bile acids. Plasma of control subjects showed increased levels of the above mentioned parameters with after incubation for 1 hour and addition of cholesterol substrate resulted in still further significant increase in these parameters. The plasma of patients showed similar results but the extent of increase was more.

The addition of antibiotics to the control plasma caused a decrease in all the parameters while addition of rutile increased their levels. The addition of antibiotics to the patient's plasma caused a decrease in all the parameters while addition of rutile increased their levels but the extent of change was more in patient's sera as compared to controls. The results are expressed in tables 1-7 as percentage change in the parameters after 1 hour incubation as compared to the values at zero time.

Discussion

There was increase in cytochrome F420 indicating archaeal growth in multiple sclerosis , systemic lupus erythematosis and rheumatoid arthritis. The archaea can synthesise and use cholesterol as a carbon and energy source [14,15]. The archaeal origin of the enzyme activities was indicated by antibiotic induced suppression. The study indicates the presence of actinide based archaea with an alternate actinide based enzymes or metalloenzymes in the system as indicated by rutile induced increase in enzyme activities [16]. There was also an increase in archaeal HMG CoA reductase activity indicating increased cholesterol synthesis by the archaeal mevalonate pathway. The archaeal beta hydroxyl steroid dehydrogenase activity indicating digoxin synthesis and archaeal cholesterol hydroxylase activity indicating bile acid synthesis were increased [7]. The archaeal cholesterol oxidase activity was increased resulting in generation of pyruvate and hydrogen peroxide [15]. The pyruvate gets converted to glutamate and ammonia by the GABA shunt pathway. The archaeal aromatization of cholesterol generating PAH, serotonin and dopamine was also detected [17]. The archaeal glycolytic hexokinase activity and archaeal extracellular ATP synthase activity were increased.

The archaea can undergo magnetite and calcium carbonate mineralization and can exist as calcified nanoforms [18]. There was an increase in free RNA indicating self replicating RNA viroids and free DNA indicating generation of viroid complementary DNA strands by archaeal reverse transcriptase activity. The actinides modulate RNA folding and catalyse its ribozymal action. Digoxin can cut and paste the viroidal strands by modulating RNA splicing generating RNA viroidal diversity. The viroids are evolutionarily escaped archaeal group I introns which have retrotransposition and self splicing qualities [19]. Archaeal pyruvate can produce histone deacetylase inhibition resulting in endogenous retroviral (HERV) reverse transcriptase and integrase expression. This can integrate the RNA viroidal complementary DNA into the noncoding region of eukaryotic non coding DNA using HERV integrase as has been described for borna and ebola viruses [20]. The noncoding DNA is lengthened by integrating RNA viroidal complementary DNA with the integration going on as a continuing event. The archaea genome can also get integrated into human genome using integrase as has been described for trypanosomes [21]. The integrated viroids and archaea can undergo vertical transmission and can exist as genomic parasites [20,21]. This increases the length and alters the grammar of the noncoding region producing memes or memory of acquired characters [22]. The viroidal complementary DNA can function as jumping genes producing a dynamic genome important in HLA gene expression. This modulation of HLA gene expression by viroidal complementary DNA can result in autoimmune diseases. The RNA viroids can regulate mrna function by RNA interference [19]. The phenomena of RNA interference can modulate T cell and B cell function and euchromatin/ heterochromatin expression. RNA viroidal mRNA interference plays a role in the pathogenesis of autoimmune diseases like systemic lupus erythematosis, rheumatoid arthritis and multiple sclerosis.

The presence of muramic acid, HMG CoA reductase and cholesterol oxidase activity inhibited by antibiotics indicates the presence of bacteria with mevalonate pathway. The bacterial with mevalonate pathway include streptococcus, staphylococcus, actinomycetes, listeria, coxiella and borrelia [23]. The bacteria and archaea with mevalonate pathway and cholesterol catabolism had a evolutionarily advantage and constitutes the isoprenoidal clade organism with the archaea evolving into mevalonate pathway gram positive and gram negative organism through horizontal gene transfer of viroidal and virus genes [24].

The isoprenoidal clade prokaryotes develop into other groups of prokaryotes via viroidal/virus as well as eukaryotic horizontal gene transfer producing bacterial speciation [25].

Table 1. Effect of rutile and antibiotics on ATP synthase and cytochrome F 420

	ATP synthase % (Increase with Rutile)		ATP synthase % (Decrease with Doxy)		CYT F420 % (Increase with Rutile)		CYT F420 % (Decrease with Doxy)	
Group	Mean	± SD	Mean	± SD	Mean	± SD	Mean	± SD
Normal	4.40	0.11	18.78	0.11	4.48	0.15	18.24	0.66
MS	23.52	1.76	67.05	3.00	22.12	1.81	61.33	9.82
SLE	23.04	1.66	66.13	3.49	21.65	2.25	61.19	7.90
RA	23.29	1.98	67.46	3.96	22.22	1.76	53.92	7.12
F value	449.503		673.081		306.749		130.054	
P value	< 0.001		< 0.001		< 0.001		< 0.001	

Table 2. Effect of rutile and antibiotics on free DNA and RNA

	DNA % change (Increase with Rutile)		DNA % change (Decrease with Doxy)		RNA % change (Increase with Rutile)		RNA % change (Decrease with Doxy)	
Group	Mean	± SD	Mean	± SD	Mean	± SD	Mean	± SD
Normal	4.37	0.15	18.39	0.38	4.37	0.13	18.38	0.48
MS	22.62	1.38	63.82	5.53	23.29	1.98	67.46	3.96
SLE	21.94	2.03	64.29	5.35	23.75	1.81	66.49	4.11
RA	21.93	2.29	63.70	5.63	23.52	1.76	67.05	3.00
F value	337.577		356.621		427.828		654.453	
P value	< 0.001		< 0.001		< 0.001		< 0.001	

Table 3. Effect of rutile and antibiotics on HMG CoA reductase and PAH

	HMG CoA R % change (Increase with Rutile)		HMG CoA R % change (Decrease with Doxy)		PAH % change (Increase with Rutile)		PAH % change (Decrease with Doxy)	
Group	Mean	± SD	Mean	± SD	Mean	± SD	Mean	± SD
Normal	4.30	0.20	18.35	0.35	4.45	0.14	18.25	0.72
MS	23.14	1.85	59.76	4.82	22.83	1.78	59.84	7.62
SLE	22.64	1.99	64.93	7.62	22.91	2.13	61.96	6.10
RA	23.42	1.76	65.71	3.28	22.26	1.90	61.80	6.12
F value	319.332		199.553		391.318		257.996	
P value	< 0.001		< 0.001		< 0.001		< 0.001	

Table 4. Effect o[illegible] rutile and antibiotics on digoxin and bile acids

	Digoxin (ng/ml) (Increase with Rutile)		Digoxin (ng/ml) (Decrease with Doxy+Cipro)		Bile acids % change (Increase with Rutile)		Bile acids % change (Decrease with Doxy)	
Group	Mean	± SD	Mean	± SD	Mean	± SD	Mean	± SD
Normal	0.11	0.00	0.[illegible]54	0.003	4.29	0.18	18.15	0.58
MS	0.52	0.03	0.[illegible]14	0.032	21.95	2.11	65.46	5.79
SLE	0.54	0.05	0.[illegible]27	0.036	22.88	1.99	63.92	6.74
RA	0.50	0.06	0.[illegible]28	0.027	23.35	2.05	67.25	4.72
F value	135.116		71[illegible]706		290.441		203.651	
P value	< 0.001		< [illegible].001		< 0.001		< 0.001	

Table 5. Effect of [illegible]rutile and antibiotics on pyruvate and hexokinase

	Pyruvate % change (Increase with Rutile)		Pyruvate % change (Decrease with Doxy)		Hexokinase % change (Increase with Rutile)		Hexokinase % change (Decrease with Doxy)	
Group	Mean	± SD	Mean	± SD	Mean	± SD	Mean	± SD
Normal	4.34	0.21	1[illegible]43	0.82	4.21	0.16	18.56	0.76
MS	21.59	1.23	6[illegible].28	9.22	22.81	1.91	63.47	5.81
SLE	20.01	1.73	5[illegible].08	8.16	21.84	2.08	63.56	5.23
RA	20.08	1.84	6[illegible].70	7.17	22.65	1.99	64.05	4.89
F value	321.255		1[illegible]5.242		292.065		317.966	
P value	< 0.001		< 0.001		< 0.001		< 0.001	

Table 6. Effect of rutile and antibiotics on hydrogen peroxide and delta amino levulinic acid

	H_2O_2 % (Increase with Rutile)		H_2O_2 % (Decrease with Doxy)		ALA % (Increase with Rutile)		ALA % (Decrease with Doxy)	
Group	Mean	± SD	Mean	± SD	Mean	± SD	Mean	± SD
Normal	4.43	0.19	18.13	0.63	4.40	0.10	18.48	0.39
MS	21.14	1.20	50.53	4.70	22.38	1.79	67.10	3.82
SLE	23.71	1.64	59.51	6.63	23.89	1.69	65.09	3.89
RA	23.03	1.66	61.91	6.31	22.92	2.13	67.54	3.65
F value	380.721		171.228		372.716		556.411	
P value	< 0.001		< 0.001		< 0.001		< 0.001	

Table 7. Effect of rutile and antibiotics on muramic acid and serotonin

	Muramic acid % change (Increase with Rutile)		Muramic acid % change (Decrease with Doxy+Cipro)		5 HT % (Increase without Doxy)		5 HT % (Decrease with Doxy)	
Group	Mean	± SD	Mean	± SD	Mean	± SD	Mean	± SD
Normal	4.41	0.15	18.63	0.12	4.34	0.15	18.24	0.37
MS	22.92	2.14	67.54	3.65	21.93	2.29	63.70	5.63
SLE	23.06	1.72	64.82	3.31	23.11	2.00	61.52	4.97
RA	22.38	1.79	67.10	3.82	22.62	1.38	63.82	5.53
F value	403.394		680.284		348.867		364.999	
P value	< 0.001		< 0.001		< 0.001		< 0.001	

The RNA viroids and its complementary DNA developed into cholesterol enveloped RNA and DNA viruses like herpes, retrovirus, influenza virus, borna virus, cytomegalo virus and Ebstein Barr virus by recombining with eukaryotic and human genes resulting in viral speciation. Bacterial and viral species are ill defined and fuzzy with all of them forming one common genetic pool with frequent horizontal gene transfer and recombination. Thus the multi and unicellular eukaryote with its genes serves the purpose of prokaryotic and viral speciation. The multicellular eukaryote developed so that their endosymbiotic archaeal colonies could survive and forage better. The multicellular eukaryotes are like bacterial biofilms. The archaea and bacteria with a mevalonate pathway uses the extracellular RNA viroids and DNA viroids for quorum sensing and in the generation of symbiotic biofilm like structures which develop into multicellular eukaryotes [26,27]. The endosymbiotic archaea and bacteria with mevalonate pathway still uses the RNA viroids and DNA viroids for the regulation of muticellular eukaryote. Pollution is induced by the primitive nanoarchaea and mevalonate pathway bacteria synthesised PAH and methane leading on to redox stress. Redox stress leads to sodium potassium ATPase inhibition, inward movement of plasma membrane cholesterol, defective SREBP sensing, increased cholesterol synthesis and nanoarchaeal/mevalonate pathway bacterial growth [28]. Redox stress leads on to viroidal and archaeal multiplication. Redox stress can also lead to HERV reverse transcriptase and integrase expression. The noncoding DNA is formed of integrating RNA viroidal complementary DNA and archaea with the integration going on as a continuing event. The archaeal pox like dsDNA virus forms evolutionarily the nucleus. The integrated viroidal, archaeal and mevalonate pathway bacterial sequences can undergo vertical transmission and can exist as genomic parasites. The genomic integrated archaea, mevalonate pathway bacteria and viroids form a genomic reserve of bacteria and viruses which can recombine with human and eukaryotic genes producing bacterial and viral speciation. Bacteria and viruses have been related to the pathogenesis of systemic lupus erythematosis, rheumatoid arthritis and multiple sclerosis. Mollicutes, streptococcus, mycoplasma, ebstein barr virus and herpes virus have been related to the pathogenesis of SLE [29]. Mycoplasma, borrelia, chlamydia, rickettsia, streptococcus, ebstein barr virus and herpes viruses have been related to the evolution of MS [30,31,32]. Mycoplasma, proteus, mollicutes, streptococcus, hepatitis C, arbo virus and parvo virus have been implicated in the pathogenesis of rheumatoid arthritis [33,34,35]. The change in the length and grammar of the noncoding region produces eukaryotic speciation and individuality [36]. Changes in the length of noncoding region especially human endogenous retroviruses can lead onto autoimmune diseases like SLE, MS and RA [37]. The integration of nanoarchaea, mevalonate pathway prokaryotes and viroids in to the eukaryotic and human genome produces a chimera which can multiply producing biofilm like multicellular structures having a mixed archaeal, viroidal, prokaryotic and eukaryotic characters which is a regression from the multicellular eukaryotic tissue This results in a new neuronal, metabolic, immune and tissue phenotype leading to human diseases like systemic lupus erythematosis, rheumatoid arthritis and multiple sclerosis. The microchimeras formed can lead to autoantigens and autoimmune diseases.

Archaea and RNA viroid can bind the TLR receptor induce NFKB producing immune activation and cytokine TNF alpha secretion. The archaeal DXP and mevalonate pathway metabolites can bind γδ TCR and digoxin induced calcium signaling can activate NFKB producing chronic immune activation [2,38]. The archaea and viroid induced chronic immune activation and generation of superantigens can lead on to autoimmune disease. The archaea and viroids can regulate the nervous system including the NMDA synaptic transmission [2]. NMDA can be activated by digoxin induced calcium oscillations, PAH and viroid induced RNA interference [2]. The cholesterol ring oxidase generated pyruvate can be converted by the GABA shunt pathway to glutamate. The archaeal cholesterol aromatase can generate serotonin [17]. Glutamatergic and serotoninergic transmission can lead to immune activation. The higher degree of integration of the archaea into the genome produces increased digoxin synthesis producing right hemispheric dominance and lesser degree producing left hemispheric dominance [2]. Right hemispheric dominance can lead to autoimmune diseases. Archaea, viroids and digoxin can induce the host AKT PI3K, AMPK, HIF alpha and NFKB producing the Warburg metabolic phenotype [39]. The increased glycolytic hexokinase activity, decrease in blood ATP, leakage of cytochrome C, increase in serum pyruvate and decrease in acetyl CoA indicates the generation of the Warburg phenotype. There is induction of glycolysis, inhibition of PDH activity and mitochondrial dysfunction resulting in inefficient energetics. The lymphocytes depend on glycolysis for their energy needs. The increased glycolysis induced by the Warburg phenotype leads to immune activation. Lactic acid generated by increased glycolysis leads to immune stimulation. Cholesterol oxidase activity, increased glycolysis related NADPH oxidase activity and mitochondrial dysfunction generates free radicals important in the pathogenesis of these disease states. The accumulated pyruvate enters the gaba shunt pathway and is converted to citrate which is acted upon by citrate lyase and converted to acetyl CoA, used for cholesterol synthesis [39]. The pyruvate can be converted to glutamate and ammonia which is oxidised by archaea for energy needs. The increased cholesterol substrate leads to increased archaeal growth and digoxin synthesis leading to metabolic channeling to the mevalonate pathway. Hyperdigoxinemia is important in the pathogenesis of autoimmune disease. Digoxin can increase lymphocytic intracellular calcium which leads on to induction of NFKB and immune activation [2]. The archaeal cholesterol catabolism can deplete the lymphocytic cell membranes of cholesterol resulting in alteration of lymphocytic cell membrane microdomains related receptors producing immune activation. Thus the actinide, viroid and mevalonate pathway bacteria induced metabolic, genetic, immune and neuronal transmission changes can lead onto autoimmune diseases.

References

[1] Valiathan M.S., Somers, K., Kartha, C.C. (1993). *Endomyocardial Fibrosis.* Delhi: Oxford University Press.

[2] Kurup R., Kurup, P.A. (2009). *Hypothalamic digoxin, cerebral dominance and brain function in health and diseases.* New York: Nova Science Publishers.

[3] Hanold D., Randies, J.W. (1991). Coconut cadang-cadang disease and its viroid agent, *Plant Disease,* 75, 330-335.

[4] Edwin B.T., Mohankumaran, C. (2007). Kerala wilt disease phytoplasma: Phylogenetic analysis and identification of a vector, *Proutista moesta, Physiological and Molecular Plant Pathology,* 71(1-3), 41-47.

[5] Eckburg P.B., Lepp, P.W., Relman, D.A. (2003). Archaea and their potential role in human disease, *Infect. Immun.,* 71, 591-596.

[6] Adam Z. (2007). Actinides and Life's Origins, *Astrobiology,* 7, 6-10.

[7] Schoner W. (2002). Endogenous cardiac glycosides, a new class of steroid hormones, *Eur. J. Biochem.,* 269, 2440-2448.

[8] Davies P.C.W., Benner, S.A., Cleland, C.E., Lineweaver, C.H., McKay, C.P., Wolfe-Simon, F. (2009). Signatures of a Shadow Biosphere, *Astrobiology,* 10, 241-249.

[9] Richmond W. (1973). Preparation and properties of a cholesterol oxidase from nocardia species and its application to the enzymatic assay of total cholesterol in serum, *Clin. Chem.,* 19, 1350-1356.

[10] Snell E.D., Snell, C.T. (1961). *Colorimetric Methods of Analysis.* Vol 3A. New York: Van NoStrand.

[11] Glick D. (1971). *Methods of Biochemical Analysis.* Vol 5. New York: Interscience Publishers.

[12] Colowick, Kaplan, N.O. (1955). *Methods in Enzymology.* Vol 2. New York: Academic Press.

[13] Maarten A.H., Marie-Jose, M., Cornelia, G., van Helden–Meewsen, Fritz, E., Marten, P.H. (1995). Detection of muramic acid in human spleen, *Infection and Immunity,* 63(5), 1652 – 1657.

[14] Smit A., Mushegian, A. (2000). Biosynthesis of isoprenoids via mevalonate in Archaea: the lost pathway, *Genome Res.,* 10(10), 1468-84.

[15] Van der Geize R., Yam, K., Heuser, T., Wilbrink, M.H., Hara, H., Anderton, M.C. (2007). A gene cluster encoding cholesterol catabolism in a soil actinomycete provides insight into Mycobacterium tuberculosis survival in macrophages, *Proc. Natl. Acad. Sci. USA,* 104(6), 1947-52.

[16] Francis A.J. (1998). Biotransformation of uranium and other actinides in radioactive wastes, *Journal of Alloys and Compounds,* 271(273), 78-84.

[17] Probian C., Wülfing, A., Harder, J. (2003). Anaerobic mineralization of quaternary carbon atoms: Isolation of denitrifying bacteria on pivalic acid (2,2-Dimethylpropionic acid), *Applied and Environmental Microbiology,* 69(3), 1866-1870.

[18] Vainshtein M., Suzina, N., Kudryashova, E., Ariskina, E. (2002). New Magnet-Sensitive Structures in Bacterial and Archaeal Cells, *Biol. Cell,* 94(1), 29-35.

[19] Tsagris E.M., de Alba, A.E., Gozmanova, M., Kalantidis, K. (2008). Viroids, *Cell Microbiol,* 10, 2168.

[20] Horie M., Honda, T., Suzuki, Y., Kobayashi, Y., Daito, T., Oshida, T. (2010). Endogenous non-retroviral RNA virus elements in mammalian genomes, *Nature,* 463, 84-87.

[21] Hecht M., Nitz, N., Araujo, P., Sousa, A., Rosa, A., Gomes, D. (2010). Genes from Chagas parasite can transfer to humans and be passed on to children. Inheritance of DNA Transferred from American Trypanosomes to Human Hosts, *PLoS ONE,* 5, 2-10.

[22] Flam F. (1994). Hints of a language in junk DNA, *Science,* 266, 1320.

[23] Horbach S., Sahm, H., Welle, R. (1993). Isoprenoid biosynthesis in bacteria: two different pathways? *FEMS Microbiol. Lett,* 111, 135–140.

[24] Gupta R.S. (1998). Protein phylogenetics and signature sequences: a reappraisal of evolutionary relationship among archaebacteria, eubacteria, and eukaryotes, *Microbiol. Mol. Biol. Rev.,* 62, 1435–1491.

[25] Hanage W., Fraser, C., Spratt, B. (2005). Fuzzy species among recombinogenic bacteria, *BMC Biology,* 3, 6-10.

[26] Webb J.S., Givskov, M., Kjelleberg, S. (2003). Bacterial biofilms: prokaryotic adventures in multicellularity, *Curr. Opin. Microbiol,* 6(6), 578–85.

[27] Whitchurch C.B., Tolker-Nielsen, T., Ragas, P.C., Mattick, J.S. (2002). Extracellular DNA Required for Bacterial Biofilm Formation. *Science,* 295(5559), 1487.

[28] Chen Y., Cai, T., Wang, H., Li, Z., Loreaux, E., Lingrel, J.B. (2009). Regulation of intracellular cholesterol distribution by Na/K-ATPase, *J. Biol. Chem.,* 284(22), 14881-90.

[29] Krieg A.M. (1995). CpG DNA: a pathogenic factor in systemic lupus erythematosus? *J. Clin. Immunol.,* 15(6), 284-92.

[30] Schmutzhard E., Pohl, P., Stanek, G. (1988). Borrelia burgdorferi antibodies in patients with relapsing/remitting form and chronic progressive form of multiple sclerosis, *J. Neurol. Neurosurg. Psychiatry,* 51(9), 1215-8.

[31] Sriram S., Stratton, C.W., Yao, S., Tharp, A., Ding, L., Bannan, J.D., Mitchell, W.M. (1999). Chlamydia Pneumoniae Infection of the Central Nervous System in Multiple Sclerosis, *Ann. Neurol.,* 46(1), 6-14.

[32] Maida E. (1983). Immunological reactions against Mycoplasma pneumoniac in multiple sclerosis: preliminary findings, *Journal of Neurology,* 229(2), 103-111.

[33] Vojdani A., Franco, A.R. (1999). Multiplex PCR for the detection of *M. fermentans, M. hominis* and *M. penetrans* in patients with Chronic Fatigue Syndrome, Fibromyalgia, Rheumatoid Arthritis and Gulf War Syndrome, *J. Chron. Fatig,* 5(3/4), 187-197.

[34] Wilson C., Thakore, A., Isenberg, D., Ebringer, A. (1997). Correlation between anti-Proteus antibodies and isolation rates of P-mirabilis in rheumatoid arthritis, *Rheumatology International,* 16(5), 187-189.

[35] Wirostko E., Johnson, L., Wirostko, W. (1989). Juvenile rheumatoid arthritis inflammatory eye disease. Parasitization of ocular leukocytes by mollicute-like organisms, *J. Rheumatol.,* 16(11), 1446–1453.

[36] Poole A.M. (2006). Did group II intron proliferation in an endosymbiont-bearing archaeon create eukaryotes? *Biol. Direct,* 1, 36-40.

[37] Villarreal L.P. (2006). How viruses shape the tree of life, *Future Virology,* 1(5), 587-595.

[38] Eberl M., Hintz, M., Reichenberg, A., Kollas, A., Wiesner, J., Jomaa, H. (2010). Microbial isoprenoid biosynthesis and human γδ T cell activation, *FEBS Letters,* 544(1), 4-10.

[39] Wallace D.C. (2005). Mitochondria and Cancer: Warburg Addressed, *Cold Spring Harbor Symposia on Quantitative Biology,* 70, 363-374.

Chapter XII

A Cholesterol and Actinide Dependent Shadow Biosphere of Archaea and Viroids in Neoplastic Disorders

Introduction

The human body secretes an endogenous sodium-potassium ATPase inhibitor, digoxin which can regulate the cell cycle and cell proliferation. Endomyocardial fibrosis (EMF) along with the root wilt disease of coconut is endemic to Kerala with its radioactive actinide beach sands. Actinides like rutile producing intracellular magnesium deficiency due to rutile-magnesium exchange sites in the cell membrane has been implicated in the etiology of EMF [1]. Endogenous digoxin has also been related to its etiology due to the intracellular magnesium deficiency it produces [2]. Organisms like phytoplasmas and viroids have also been demonstrated to play a role in the etiology of these diseases [3,4]. Endogenous digoxin has been related to the pathogenesis of non-hodgkin's B cell lymphoma, multiple myeloma and CNS glioma- glioblastoma multiforme [2]. The incidence of neoplasms is high in the presence of low level radioactivity of the mineral sands of Kerala [1]. The possibility of endogenous digoxin synthesis by actinide based primitive organism like archaea with a mevalonate pathway and cholesterol catabolism was considered [5,6,7]. Davies has put forward the concept of a shadow biosphere of organisms with alternate biochemistry present in earth itself [8]. An actinide dependent shadow biosphere of archaea and viroids in the above mentioned disease states is described [6].

Materials and Methods

Informed consent of the subjects and the approval of the ethics committee were obtained for the study. The following groups were included in the study:- non-hodgkin's B cell lymphoma, multiple myeloma and cns glioma- glioblastoma multiforme. There were 10 patients in each group and each patient had an age and sex matched healthy control selected randomly from the general population. The blood samples were drawn in the fasting state before treatment was initiated. Plasma from fasting heparinised blood was used and the experimental protocol was as follows (I) Plasma+phosphate buffered saline, (II) same as I+cholesterol substrate, (III) same as II+rutile 0.1 mg/ml, (IV) same as II+ciprofloxacine and doxycycline each in a concentration of 1 mg/ml. Cholesterol substrate was prepared as described by Richmond [9]. Aliquots were withdrawn at zero time immediately after mixing and after incubation at 37°C for 1 hour. The following estimations were carried out:- Cytochrome F420, free RNA, free DNA, muramic acid, polycyclic aromatic hydrocarbon, hydrogen peroxide, serotonin, pyruvate, ammonia, glutamate, cytochrome C, hexokinase, ATP synthase, HMG CoA redutase, digoxin and bile acids [10,11,12,13]. Cytochrome F420 was estimated flourimetrically (excitation wavelength 420 nm and emission wavelength 520 nm). Polycyclic aromatic hydrocarbon was estimated by measuring hydrogen peroxide liberated by using glucose reagent. The statistical analysis was done by ANOVA.

Results

The parameters checked as indicated above were:- cytochrome F420, free RNA, free DNA, muramic acid, polycyclic aromatic hydrocarbon, hydrogen peroxide, serotonin, pyruvate, ammonia, glutamate, cytochrome C, hexokinase, ATP synthase, HMG CoA redutase, digoxin and bile acids. Plasma of control subjects showed increased levels of the above mentioned parameters with after incubation for 1 hour and addition of cholesterol substrate resulted in still further significant increase in these parameters. The plasma of patients showed similar results but the extent of increase was more. The addition of antibiotics to the control plasma caused a decrease in all the parameters while addition of rutile increased their levels. The addition of antibiotics to the patient's plasma caused a decrease in all the parameters while addition of rutile increased their levels but the extent of change was more in patient's sera as compared to controls. The results are expressed in tables 1-7 as percentage change in the parameters after 1 hour incubation as compared to the values at zero time.

Discussion

There was increase in cytochrome F420 indicating archaeal growth in non-hodgkin's lymphoma- diffuse large B cell type, multiple myeloma and cns glioma-glioblastoma multiforme. The archaea can synthesise and use cholesterol as a carbon and energy source [14,15]. The archaeal origin of the enzyme activities was indicated by antibiotic induced suppression. The study indicates the presence of actinide based archaea with an alternate actinide based enzymes or metalloenzymes in the system as indicated by rutile induced increase in enzyme activities [16]. There was also an increase in archaeal HMG CoA reductase activity indicating increased cholesterol synthesis by the archaeal mevalonate pathway. The archaeal beta hydroxyl steroid dehydrogenase activity indicating digoxin synthesis and archaeal cholesterol hydroxylase activity indicating bile acid synthesis were increased [7]. The archaeal cholesterol oxidase activity was increased resulting in generation of pyruvate and hydrogen peroxide [15].

Table 1. Effect of rutile and antibiotics on muramic acid and serotonin

	Muramic acid % (Increase without Doxy)		Muramic acid % (Decrease with Doxy)		5 HT % (Increase without Doxy)		5 HT % (Decrease with Doxy)	
Group	Mean	± SD	Mean	± SD	Mean	± SD	Mean	± SD
Normal	4.41	0.15	[illegible]8.63	0.12	4.34	0.15	18.24	0.37
Myel	22.87	1.84	[illegible]6.31	3.68	23.01	1.67	65.35	3.56
NHL	23.81	1.90	[illegible]6.95	3.67	23.12	1.71	65.12	5.58
Glio	22.63	1.63	[illegible]7.24	3.42	22.51	1.85	63.56	5.29
F value	403.394		[illegible]80.284		348.867		364.999	
P value	< 0.001		< 0.001		< 0.001		< 0.001	

Table 2. Effect of rutile and antibiotics on free DNA and RNA

	DNA % change (Increase with Rutile)		DNA % change (Decrease with Doxy)		RNA % change (Increase with Rutile)		RNA % change (Decrease with Doxy)	
Group	Mean	± SD	Mean	± SD	Mean	± SD	Mean	± SD
Normal	4.37	0.15	18.39	0.38	4.37	0.13	18.38	0.48
Myel	22.73	2.46	65.87	4.35	23.72	1.73	66.25	3.69
NHL	22.42	1.99	61.14	3.47	23.78	1.20	66.90	4.10
Glio	22.92	1.99	66.55	4.55	23.47	1.60	66.27	3.88
F value	337.577		356.621		427.828		654.453	
P value	< 0.001		< 0.001		< 0.001		< 0.001	

Table 3. Effect of rutile and antibiotics on HMG CoA reductase and PAH

	HMG CoA R % change (Increase with Rutile)		HMG CoA R % change (Decrease with Doxy)		PAH % change (Increase with Rutile)		PAH % change (Decrease with Doxy)	
Group	Mean	± SD	Mean	± SD	Mean	± SD	Mean	± SD
Normal	4.30	0.20	18.35	0.35	4.45	0.14	18.25	0.72
Myel	23.54	1.79	61.75	5.76	23.60	2.00	62.37	5.58
NHL	22.28	1.76	61.88	6.21	22.84	1.42	66.07	3.78
Glio	23.18	2.09	63.87	5.36	22.87	1.59	62.02	6.89
F value	319.332		199.553		391.318		257.996	
P value	< 0.001		< 0.001		< 0.001		< 0.001	

Table 4. Effect of rutile and antibiotics on digoxin and bile acids

	Digoxin (ng/ml) (Increase with Rutile)		Digoxin (ng/ml) (Decrease with Doxy+Cipro)		Bile acids % change (Increase with Rutile)		Bile acids % change (Decrease with Doxy)	
Group	Mean	± SD	Mean	± SD	Mean	± SD	Mean	± SD
Normal	0.11	0.00	0.054	0.003	4.29	0.18	18.15	0.58
Myel	0.50	0.03	0.208	0.039	22.05	1.69	62.73	7.31
NHL	0.54	0.04	0.210	0.042	22.98	2.19	64.96	5.64
Glio	0.50	0.05	0.195	0.026	23.31	2.05	61.67	4.54
F value	135.116		71.706		290.441		203.651	
P value	< 0.001		< 0.001		< 0.001		< 0.001	

Table 5. Effect of rutile and antibiotics on pyruvate and hexokinase

	Pyruvate % change (Increase with Rutile)		Pyruvate % change (Decrease with Doxy)		Hexokinase % change (Increase with Rutile)		Hexokinase % change (Decrease with Doxy)	
Group	Mean	± SD	Mean	± SD	Mean	± SD	Mean	± SD
Normal	4.34	0.21	18.43	0.82	4.21	0.16	18.56	0.76
Myel	20.01	2.11	61.13	5.37	22.29	1.98	64.44	6.24
NHL	21.19	1.61	58.57	7.47	22.53	2.41	64.29	5.44
Glio	20.79	1.65	53.57	5.89	21.82	1.86	64.26	6.05
F value	321.255		115.242		292.065		317.966	
P value	< 0.001		< 0.001		< 0.001		< 0.001	

Table 6. Effect of rutile and antibiotics on hydrogen peroxide and delta amino levulinic acid

	H_2O_2 % (Increase with Rutile)		H_2O_2 % (Decrease with Doxy)		ALA % (Increase with Rutile)		ALA % (Decrease with Doxy)	
Group	Mean	± SD	Mean	± SD	Mean	± SD	Mean	± SD
Normal	4.43	0.19	18.13	0.63	4.40	0.10	18.48	0.39
Myel	23.35	1.95	58.75	3.75	24.10	1.61	65.78	4.43
NHL	23.35	1.76	59.17	3.33	23.34	1.75	66.80	3.43
Glio	22.71	1.73	57.49	8.26	22.95	1.61	65.76	4.01
F value	380.721		171.228		372.716		556.411	
P value	< 0.001		< 0.001		< 0.001		< 0.001	

Table 7. Effect of rutile and antibiotics on ATP synthase and cytochrome F 420

	ATP synthase % (Increase with Rutile)		ATP synthase % (Decrease with Doxy)		CYT F420 % (Increase with Rutile)		CYT F420 % (Decrease with Doxy)	
Group	Mean	± SD	Mean	± SD	Mean	± SD	Mean	± SD
Normal	4.40	0.11	8.78	0.11	4.48	0.15	18.24	0.66
Myel	23.33	1.86	[illegible]6.46	3.65	23.01	1.89	61.17	9.61
NHL	24.01	1.17	56.66	3.84	22.79	2.13	55.90	7.29
Glio	22.95	1.90	56.39	3.83	22.03	1.64	62.21	5.53
F value	449.503		573.081		306.749		130.054	
P value	< 0.001		< 0.001		< 0.001		< 0.001	

The pyruvate gets converted to glutamate and ammonia by the GABA shunt pathway. The archaeal aromatization of cholesterol generating PAH, serotonin and dopamine was also detected [17]. The archaeal glycolytic hexokinase activity and archaeal extracellular ATP synthase activity were increased. The archaea can undergo magnetite and calcium carbonate mineralization and can exist as calcified nanoforms [18]. There was an increase in free RNA indicating self replicating RNA viroids and free DNA indicating generation of viroid complementary DNA strands by archaeal reverse transcriptase activity. The actinides modulate RNA folding and catalyse its ribozymal action.

Digoxin can cut and paste the viroidal strands by modulating RNA splicing generating RNA viroidal diversity. The viroids are evolutionarily escaped archaeal group I introns which have retrotransposition and self splicing qualities [19]. Archaeal pyruvate can produce histone deacetylase inhibition resulting in endogenous retroviral (HERV) reverse transcriptase and integrase expression. This can integrate the RNA viroidal complementary DNA into the noncoding region of eukaryotic non coding DNA using HERV integrase as has been described for borna and ebola viruses [20]. The noncoding DNA is lengthened by integrating RNA viroidal complementary DNA with the integration going on as a continuing event. The archaea genome can also get integrated into human genome using integrase as has been described for trypanosomes [21]. The integrated viroids and archaea can undergo vertical transmission and can exist as genomic parasites [20,21]. This increases the length and alters the grammar of the noncoding region producing memes or memory of acquired characters [22]. The viroidal complementary DNA can function as jumping genes producing a dynamic genome modulating gene expression. The RNA viroids can regulate mrna function by RNA interference [19]. The phenomena of RNA interference can modulate T cell and B cell function, cell differentiation, cell growth, apoptosis and euchromatin/heterochromatin expression. RNA viroidal mRNA interference can modulate the cell cycle producing malignant transformation.

The presence of muramic acid, HMG CoA reductase and cholesterol oxidase activity inhibited by antibiotics indicates the presence of bacteria with mevalonate pathway. The bacterial with mevalonate pathway include streptococcus, staphylococcus, actinomycetes, listeria, coxiella and borrelia [23]. The bacteria and archaea with mevalonate pathway and cholesterol catabolism had a evolutionarily advantage and constitutes the isoprenoidal clade organism with the archaea evolving into mevalonate pathway gram positive and gram negative organism through horizontal gene transfer of viroidal and virus genes [24]. The isoprenoidal clade prokaryotes develop into other groups of prokaryotes via viroidal/virus as well as eukaryotic horizontal gene transfer producing bacterial speciation [25]. The RNA viroids and its complementary DNA developed into cholesterol enveloped RNA and DNA viruses like herpes, retrovirus, influenza virus, borna virus, cytomegalo virus and Ebstein Barr virus by recombining with eukaryotic and human genes resulting in viral speciation. Bacterial and viral species are ill defined and fuzzy with all of them forming one common genetic pool with frequent horizontal gene transfer and recombination. Thus the multi and unicellular eukaryote with its genes serves the purpose of prokaryotic and viral speciation. The multicellular eukaryote developed so that their endosymbiotic archaeal colonies could survive and forage better. The multicellular eukaryotes are like bacterial biofilms. The archaea and bacteria with a mevalonate pathway uses the extracellular RNA viroids and DNA viroids for quorum sensing and in the generation of symbiotic biofilm like structures which develop into multicellular eukaryotes [26,27]. The endosymbiotic archaea and bacteria with mevalonate pathway still uses the RNA viroids and DNA viroids for the regulation of muticellular eukaryote. Pollution is induced by the primitive nanoarchaea and mevalonate pathway bacteria synthesised PAH and methane leading on to redox stress. Redox stress leads to sodium potassium ATPase inhibition, inward movement of plasma membrane cholesterol, defective SREBP sensing, increased cholesterol synthesis and nanoarchaeal/mevalonate pathway bacterial growth [28]. Redox stress leads on to viroidal and archaeal multiplication. Redox stress can also lead to HERV reverse transcriptase and integrase expression. The noncoding DNA is formed of integrating RNA viroidal complementary DNA and archaea with the integration going on as a continuing event. The archaeal pox like dsDNA virus forms evolutionarily the nucleus. The integrated viroidal, archaeal and mevalonate pathway bacterial sequences can undergo vertical transmission and can exist as genomic parasites. The genomic integrated archaea, mevalonate pathway bacteria and viroids form a genomic reserve of bacteria and viruses which can recombine with human and eukaryotic genes producing bacterial and viral speciation. Bacteria and viruses have been related to the pathogenesis of malignancy. Toll receptor activation has been related to malignant transformation. Chlamydia, mycoplasma and acid fast bacteria as well as ebstein barr virus and retroviruses have been related to the pathogenesis of lymphomas. Staphylococcus has been related to carcinoma of the breast. Herpes viruses have been related to multiple myeloma [29,30,31,32]. The change in the length and grammar of the noncoding region produces eukaryotic speciation and individuality [33]. Changes in the length of noncoding region especially human endogenous retroviruses can lead onto lymphomas, teratocarcinomas and other malignancies [34]. The integration of nanoarchaea, mevalonate pathway prokaryotes and viroids in to the eukaryotic and human genome produces a chimera which can multiply producing biofilm like multicellular structures having a mixed archaeal, viroidal, prokaryotic and eukaryotic characters which is a regression from the multicellular eukaryotic tissue This results in a new neuronal, metabolic, immune and tissue phenotype leading to malignant transformation. The microchimeras formed can lead to polyploidy and neoplastic changes.

Archaea, viroids and digoxin can induce the host AKT PI3K, AMPK, HIF alpha and NFKB producing the Warburg metabolic phenotype [35]. The increased glycolytic hexokinase activity, decrease in blood ATP, leakage of cytochrome C, increase in serum pyruvate and decrease in acetyl CoA indicates the generation of the Warburg phenotype. There is induction of glycolysis, inhibition of PDH activity and mitochondrial dysfunction resulting in inefficient energetics. The increased glycolysis results in the upregulation of mitochondrial PT pore hexokinase resulting in cell proliferation and malignant transformation. The accumulated pyruvate enters the gaba shunt pathway and is converted to citrate which is acted upon by citrate lyase and converted to acetyl CoA, used for cholesterol and lipid synthesis [35]. The increased lipid synthesis is required for membrane formation needed for cell proliferation. Cholesterol oxidase activity, increased glycolysis related NADPH oxidase activity and mitochondrial dysfunction generates free radicals important in the pathogenesis of malignancies. Free radicals can activate oncogenes producing malignant transformation. The pyruvate can be converted to glutamate and ammonia which is oxidised by archaea for energy needs. The increased cholesterol substrate also leads to increased archaeal growth and digoxin synthesis leading to metabolic channeling to the mevalonate pathway. Hyperdigoxinemia is important in the pathogenesis of tumourogenesis. Digoxin can increase lymphocytic intracellular calcium which leads on to activation of oncogenes and cell proliferation [2]. The archaeal cholesterol catabolism can deplete the lymphocytic cell membranes of cholesterol resulting in alteration of cell membrane microdomains related to oncogenic cell growth factors protein tyrosine kinase receptors and AKT PI3K producing cell proliferation and malignant transformation. Cholesterol depleted states can lead onto polyploidy and neoplastic changes.

Archaea and RNA viroid can bind the TLR receptor induce NFKB producing immune activation and cytokine TNF alpha secretion. The archaeal DXP and mevalonate pathway metabolites can bind $\gamma\delta$ TCR and digoxin induced calcium signaling can activate NFKB producing chronic immune activation [2,36]. The archaea and viroid induced chronic immune activation and generation of superantigens. NFKB can promote cell growth and proliferation leading onto malignant transformation. The archaea and viroids can regulate the nervous system including the NMDA synaptic transmission [2]. NMDA can be activated by digoxin induced calcium oscillations, PAH and viroid induced RNA interference [2]. The cholesterol ring oxidase generated pyruvate can be converted by the GABA shunt pathway to glutamate. The archaeal cholesterol aromatase can generate serotonin [17]. Glutamatergic and serotoninergic transmission can lead on to cell proliferation and malignant transformation. Mood disorders and schizophrenia can predispose to malignancies. The higher degree of integration of the archaea into the genome produces increased digoxin synthesis producing right hemispheric dominance and lesser degree producing left hemispheric dominance [2]. Right hemispheric dominance can lead to malignant transformation as reported previously from this laboratory. The archaeal cholesterol catabolism also generates PAH which can modulate gap junction intercellular communication resulting in cell proliferation and malignant transformation. Archaeal PAH can thus induce neoplastic change [37]. Archaeal bile acids tend to have an effect opposite to that of digoxin reducing free radical stress and possibly inhibiting malignant transformation. The archaeal bile acids can bind GPCR and modulate D2 regulating the conversion of T4 to T3. T3 activates uncoupling proteins reducing redox stress. Bile acids can also activate NRF ½ inducing NQO1, GST, HOI reducing redox stress. The archaeal bile acids can bind VDR, the vitamin D receptor resulting in inhibition of malignant transformation [38,39]. Thus the actinide, viroid and mevalonate pathway bacteria induced metabolic, genetic, immune and neuronal transmission changes can lead onto malignant transformation.

References

[1] Valiathan M.S., Somers, K., Kartha, C.C. (1993). *Endomyocardial Fibrosis.* Delhi: Oxford University Press.

[2] Kurup R., Kurup, P.A. (2009). *Hypothalamic digoxin, cerebral dominance and brain function in health and diseases*. New York: Nova Science Publishers.

[3] Hanold D., Randies, J.W. (1991). Coconut cadang-cadang disease and its viroid agent, *Plant Disease,* 75, 330-335.

[4] Edwin B.T., Mohankumaran, C. (2007). Kerala wilt disease phytoplasma: Phylogenetic analysis and identification of a vector, *Proutista moesta, Physiological and Molecular Plant Pathology,* 71(1-3), 41-47.

[5] Eckburg P.B., Lepp, P.W., Relman, D.A. (2003). Archaea and their potential role in human disease, *Infect Immun,* 71, 591-596.

[6] Adam Z. (2007). Actinides and Life's Origins, *Astrobiology,* 7, 6-10.

[7] Schoner W. (2002). Endogenous cardiac glycosides, a new class of steroid hormones, *Eur. J. Biochem.,* 269, 2440-2448.

[8] Davies P.C.W., Benner, S.A., Cleland, C.E., Lineweaver, C.H., McKay, C.P., Wolfe-Simon, F. (2009). Signatures of a Shadow Biosphere, *Astrobiology,* 10, 241-249.

[9] Richmond W. (1973). Preparation and properties of a cholesterol oxidase from nocardia species and its application to the enzymatic assay of total cholesterol in serum, *Clin. Chem.,* 19, 1350-1356.

[10] Snell E.D., Snell, C.T. (1961). *Colorimetric Methods of Analysis.* Vol 3A. New York: Van NoStrand.

[11] Glick D. (1971). *Methods of Biochemical Analysis.* Vol 5. New York: Interscience Publishers.

[12] Colowick, Kaplan, N.O. (1955). *Methods in Enzymology.* Vol 2. New York: Academic Press.

[13] Maarten A.H., Marie-Jose, M., Cornelia, G., van Helden–Meewsen, Fritz, E., Marten, P.H. (1995). Detection of muramic acid in human spleen, *Infection and Immunity,* 63(5), 1652 – 1657.

[14] Smit A., Mushegian, A. (2000). Biosynthesis of isoprenoids via mevalonate in Archaea: the lost pathway, *Genome Res.,* 10(10), 1468-84.

[15] Van der Geize R., Yam, K., Heuser, T., Wilbrink, M.H., Hara, H., Anderton, M.C. (2007). A gene cluster encoding cholesterol catabolism in a soil actinomycete provides insight into Mycobacterium tuberculosis survival in macrophages, *Proc. Natl. Acad. Sci. USA,* 104(6), 1947-52.

[16] Francis A.J. (1998). Biotransformation of uranium and other actinides in radioactive wastes, *Journal of Alloys and Compounds,* 271(273), 78-84.

[17] Probian C., Wülfing, A., Harder, J. (2003). Anaerobic mineralization of quaternary carbon atoms: Isolation of denitrifying bacteria on pivalic acid (2,2-Dimethylpropionic acid), *Applied and Environmental Microbiology,* 69(3), 1866-1870.

[18] Vainshtein M., Suzina, N., Kudryashova, E., Ariskina, E. (2002). New Magnet-Sensitive Structures in Bacterial and Archaeal Cells, *Biol. Cell,* 94(1), 29-35.

[19] Tsagris E.M., de Alba, A.E., Gozmanova, M., Kalantidis, K. (2008). Viroids, *Cell Microbiol.,* 10, 2168.

[20] Horie M., Honda, T., Suzuki, Y., Kobayashi, Y., Daito, T., Oshida, T. (2010). Endogenous non-retroviral RNA virus elements in mammalian genomes, *Nature,* 463, 84-87.

[21] Hecht M., Nitz, N., Araujo, P., Sousa, A., Rosa, A., Gomes, D. (2010). Genes from Chagas parasite can transfer to humans and be passed on to children. Inheritance of DNA Transferred from American Trypanosomes to Human Hosts, *PLoS ONE,* 5, 2-10.

[22] Flam F. (1994). Hints of a language in junk DNA, *Science,* 266, 1320.

[23] Horbach S., Sahm, H., Welle, R. (1993). Isoprenoid biosynthesis in bacteria: two different pathways? *FEMS Microbiol Lett.,* 111, 135–140.

[24] Gupta R.S. (1998). Protein phylogenetics and signature sequences: a reappraisal of evolutionary relationship among archaebacteria, eubacteria, and eukaryotes, *Microbiol. Mol. Biol. Rev.,* 62, 1435–1491.

[25] Hanage W., Fraser, C., Spratt, B. (2005). Fuzzy species among recombinogenic bacteria, *BMC Biology,* 3, 6-10.

[26] Webb J.S., Givskov, M., Kjelleberg, S. (2003). Bacterial biofilms: prokaryotic adventures in multicellularity, *Curr. Opin. Microbiol.,* 6(6), 578–85.

[27] Whitchurch C.B., Tolker-Nielsen, T., Ragas, P.C., Mattick, J.S. (2002). Extracellular DNA Required for Bacterial Biofilm Formation. *Science,* 295(5559), 1487.

[28] Chen Y., Cai, T., Wang, H., Li, Z., Loreaux, E., Lingrel, J.B. (2009). Regulation of intracellular cholesterol distribution by Na/K-ATPase, *J. Biol. Chem.,* 284(22), 14881-90.

[29] Tsai S., Wear, D.J., Shih, J.W.K., Lo, S.C. (1995). Mycoplasmas and oncogenesis: persistent infection and multistage malignant transformation. *Proc. Natl. Acad. Sci. USA,* 92, 10197-201.

[30] Correa P. (2000). Chemoprevention of gastric dysplasia: Randomized trial of antioxidant supplements and anti-Helicobacter pylori therapy. *Journal of the National Cancer Institute,* 92, 1881-1888.

[31] Smith J.S., Bosetti, C., Muñoz, N., Herrero, R., Bosch, F.X., Eluf-Neto, J., Meijer, C.J., Van Den Brule, A.J., Franceschi, S., Peeling, R.W. (2004). Chlamydia trachomatis and invasive cervical cancer: a pooled analysis of the IARC multicentric case-control study. *Int. J. Cancer,* 111(3), 431-9.

[32] Rakoff-nahoum S., Medzhitov, R. (2009) Toll-like receptors and cancer. *Nat Rev Cancer,* 9(1), 57.

[33] Poole A.M. (2006). Did group II intron proliferation in an endosymbiont-bearing archaeon create eukaryotes? *Biol. Direct,* 1, 36-40.

[34] Villarreal L.P. (2006). How viruses shape the tree of life, *Future Virology,* 1(5), 587-595.

[35] Wallace D.C. (2005). Mitochondria and Cancer: Warburg Addressed, *Cold Spring Harbor Symposia on Quantitative Biology,* 70, 363-374.

[36] Eberl M., Hintz, M., Reichenberg, A., Kollas, A., Wiesner, J., Jomaa, H. (2010). Microbial isoprenoid biosynthesis and human γδ T cell activation, *FEBS Letters,* 544(1), 4-10.

[37] Upham B.L., Weis, L.M., Trosko, J.E. (1998). Modulated gap junctional intercellular communication as a biomarker of PAH epigenetic toxicity: structure-function relationship. *Environ. Health Perspect,* 106(4), 975-81.

[38] Lefebvre P., Cariou, B., Lien, F., Kuipers, F., Staels, B. (2009). Role of Bile Acids and Bile Acid Receptors in Metabolic Regulation, *Physiol. Rev.,* 89(1), 147 - 191.

[39] Matsubara T, Yoshinari, K., Aoyama, K., Sugawara, M., Sekiya, Y., Nagata, K., Yamazoe, Y. (2008). Role of Vitamin D Receptor in the Lithocholic Acid-Mediated CYP3A Induction in Vitro and in Vivo. *Drug Metab. Dispos.,* 36(10), 2058 – 2063.

Chapter XIII

A Cholesterol and Actinide Dependent Shadow Biosphere of Archaea and Viroids in Retroviral and Prion Disease

Introduction

Endomyocardial fibrosis (EMF) along with the root wilt disease of coconut is endemic to Kerala with its radioactive actinide beach sands. Actinides like rutile producing intracellular magnesium deficiency due to rutile-magnesium exchange sites in the cell membrane has been implicated in the etiology of EMF [1]. Endogenous digoxin, a steroidal glycoside which functions as a membrane sodium-potassium ATPase inhibitor has also been related to its etiology due to the intracellular magnesium deficiency it produces [2]. Organisms like phytoplasmas and viroids have also been demonstrated to play a role in the etiology of these diseases [3,4]. Endogenous digoxin has been related to the pathogenesis of acquired immunodeficiency syndrome and creutzfeldt jakob disease [2]. The possibility of endogenous digoxin synthesis by actinide based primitive organism like archaea with a mevalonate pathway and cholesterol catabolism was considered [5,6,7]. Davies has put forward the concept of a shadow biosphere of organisms with alternate biochemistry present in earth itself [8]. An actinide dependent shadow biosphere of archaea and viroids in the above mentioned disease states is described [6].

Materials and Methods

Informed consent of the subjects and the approval of the ethics committee were obtained for the study. The following groups were included in the study:- acquired immunodeficiency syndrome and creutzfeldt jakob disease. There were 10 patients in each group and each patient had an age and sex matched healthy control selected randomly from the general population. The blood samples were drawn in the fasting state before treatment was initiated. Plasma from fasting heparinised blood was used and the experimental protocol was as follows (I) Plasma+phosphate buffered saline, (II) same as I+cholesterol substrate, (III) same as

II+rutile 0.1 mg/ml, (IV) same as II+ciprofloxacine and doxycycline each in a concentration of 1 mg/ml. Cholesterol substrate was prepared as described by Richmond [9]. Aliquots were withdrawn at zero time immediately after mixing and after incubation at 37°C for 1 hour. The following estimations were carried out:- Cytochrome F420, free RNA, free DNA, muramic acid, polycyclic aromatic hydrocarbon, hydrogen peroxide, serotonin, pyruvate, ammonia, glutamate, cytochrome C, hexokinase, ATP synthase, HMG CoA reductase, digoxin and bile acids [10,11,12,13]. Cytochrome F420 was estimated flourimetrically (excitation wavelength 420 nm and emission wavelength 520 nm). Polycyclic aromatic hydrocarbon was estimated by measuring hydrogen peroxide liberated by using glucose reagent. The statistical analysis was done by ANOVA.

Results

The parameters checked as indicated above were:- cytochrome F420, free RNA, free DNA, muramic acid, polycyclic aromatic hydrocarbon, hydrogen peroxide, serotonin, pyruvate, ammonia, glutamate, cytochrome C, hexokinase, ATP synthase, HMG CoA reductase, digoxin and bile acids. Plasma of control subjects showed increased levels of the above mentioned parameters with after incubation for 1 hour and addition of cholesterol substrate resulted in still further significant increase in these parameters. The plasma of patients showed similar results but the extent of increase was more. The addition of antibiotics to the control plasma caused a decrease in all the parameters while addition of rutile increased their levels. The addition of antibiotics to the patient's plasma caused a decrease in all the parameters while addition of rutile increased their levels but the extent of change was more in patient's sera as compared to controls. The results are expressed in tables 1-7 as percentage change in the parameters after 1 hour incubation as compared to the values at zero time.

Discussion

There was increase in cytochrome F420 indicating archaeal growth in acquired immunodeficiency syndrome and creutzfeldt jakob disease. The archaea can synthesise and use cholesterol as a carbon and energy source [14,15]. The archaeal origin of the enzyme activities was indicated by antibiotic induced suppression. The study indicates the presence of actinide based archaea with an alternate actinide based enzymes or metalloenzymes in the system as indicated by rutile induced increase in enzyme activities [16]. There was also an increase in archaeal HMG CoA reductase activity indicating increased cholesterol synthesis by the archaeal mevalonate pathway. The archaeal beta hydroxyl steroid dehydrogenase activity indicating digoxin synthesis and archaeal cholesterol hydroxylase activity indicating bile acid synthesis were increased [7]. The archaeal cholesterol oxidase activity was increased resulting in generation of pyruvate and hydrogen peroxide [15]. The pyruvate gets converted to glutamate and ammonia by the GABA shunt pathway. The archaeal aromatization of cholesterol generating PAH, serotonin and dopamine was also detected [17].

Table 1. Effect of rutile and antibiotics on muramic acid and serotonin

	Muramic acid % change (Increase with Rutile)		Muramic acid % change (Decrease with Doxy+Cipro)		5 HT % (Increase without Doxy)		5 HT % (Decrease with Doxy)	
Group	Mean	± SD	Mean	± SD	Mean	± SD	Mean	± SD
Normal	4.41	0.15	18.63	0.12	4.34	0.15	18.24	0.37
AIDS	23.43	1.57	66.30	3.57	22.98	1.50	65.13	4.87
CJD	23.70	1.75	68.06	3.52	23.81	1.49	64.89	6.01
F value	403.394		680.284		348.867		364.999	
P value	< 0.001		< 0.001		< 0.001		< 0.001	

Table 2. Effect of rutile and antibiotics on free DNA and RNA

	DNA % change (Increase with Rutile)		DNA % change (Decrease with Doxy)		RNA % change (Increase with Rutile)		RNA % change (Decrease with Doxy)	
Group	Mean	± SD	Mean	± SD	Mean	± SD	Mean	± SD
Normal	4.37	0.15	18.39	0.38	4.37	0.13	18.38	0.48
AIDS	22.56	2.46	62.70	4.53	23.32	1.74	65.67	4.16
CJD	23.30	1.42	65.07	4.95	23.11	1.52	66.68	3.97
F value	337.577		356.621		427.828		654.453	
P value	< 0.001		< 0.001		< 0.001		< 0.001	

Table 3. Effect of rutile and antibiotics on HMG CoA reductase and PAH

	HMG CoA R % change (Increase with Rutile)		HMG CoA R % change (Decrease with Doxy)		PAH % change (Increase with Rutile)		PAH % change (Decrease with Doxy)	
Group	Mean	± SD	Mean	± SD	Mean	± SD	Mean	± SD
Normal	4.30	0.20	18.35	0.35	4.45	0.14	18.25	0.72
AIDS	22.86	2.58	66.53	5.59	23.23	1.97	65.89	5.05
CJD	22.38	2.38	60.65	5.27	23.46	1.91	61.56	4.61
F value	319.332		199.553		391.318		257.996	
P value	< 0.001		< 0.001		< 0.001		< 0.001	

Table 4. Effect of rutile and antibiotics on digoxin and bile acids

	Digoxin (ng/ml) (Increase with Rutile)		Digoxin (ng/ml) (Decrease with Doxy+Cipro)		Bile acids % change (Increase with Rutile)		Bile acids % change (Decrease with Doxy)	
Group	Mean	± SD	Mean	± SD	Mean	± SD	Mean	± SD
Normal	0.11	0.00	0.054	0.003	4.29	0.18	18.15	0.58
AIDS	0.56	0.05	0.220	0.052	22.29	1.47	64.35	5.58
CJD	0.53	0.06	0.212	0.045	23.30	1.88	62.49	7.26
F value	135.116		71.706		290.441		203.651	
P value	< 0.001		< 0.001		< 0.001		< 0.001	

Table 5. Effect of rutile and antibiotics on pyruvate and hexokinase

	Pyruvate % change (Increase with Rutile)		Pyruvate % change (Decrease with Doxy)		Hexokinase % change (Increase with Rutile)		Hexokinase % change (Decrease with Doxy)	
Group	Mean	± SD	Mean	± SD	Mean	± SD	Mean	± SD
Normal	4.34	0.21	18.43	0.82	4.21	0.16	18.56	0.76
AIDS	21.21	2.36	58.73	8.10	21.11	2.25	64.20	5.38
CJD	21.07	1.79	63.90	7.13	22.47	2.17	65.97	4.62
F value	321.255		115.242		292.065		317.966	
P value	< 0.001		< 0.001		< 0.001		< 0.001	

Table 6. Effect of rutile and antibiotics on hydrogen peroxide and delta amino levulinic acid

	H_2O_2 % (Increase with Rutile)		H_2O_2 % (Decrease with Doxy)		ALA % (Increase with Rutile)		ALA % (Decrease with Doxy)	
Group	Mean	± SD	Mean	± SD	Mean	± SD	Mean	± SD
Normal	4.43	0.19	18.13	0.63	4.40	0.10	18.48	0.39
AIDS	23.32	1.71	63.15	7.62	23.45	1.79	66.32	3.63
CJD	22.86	1.91	63.66	6.88	23.17	1.88	68.53	2.65
F value	380.721		171.228		372.716		556.411	
P value	< 0.001		< 0.001		< 0.001		< 0.001	

Table 7. Effect of rutile and antibiotics on ATP synthase and cytochrome F 420

	ATP synthase % (Increase with Rutile)		ATP synthase % (Decrease with Doxy)		CYT F420 % (Increase with Rutile)		CYT F420 % (Decrease with Doxy)	
Group	Mean	± SD	Mean	± SD	Mean	± SD	Mean	± SD
Normal	4.40	0.11	18.78	0.11	4.48	0.15	18.24	0.66
AIDS	23.15	1.62	66.48	4.17	22.29	1.66	59.02	7.50
CJD	23.00	1.64	66.67	4.21	22.06	1.61	57.81	6.04
F value	449.503		673.081		306.749		130.054	
P value	< 0.001		< 0.001		< 0.001		< 0.001	

The archaeal glycolytic hexokinase activity and archaeal extracellular ATP synthase activity were increased. The archaea can undergo magnetite and calcium carbonate mineralization and can exist as calcified nanoforms [18]. There was an increase in free RNA indicating self replicating RNA viroids and free DNA indicating generation of viroid complementary DNA strands by archaeal reverse transcriptase activity. The actinides modulate RNA folding and catalyse its ribozymal action. Digoxin can cut and paste the viroidal strands by modulating RNA splicing generating RNA viroidal diversity. The viroids are evolutionarily escaped archaeal group I introns which have retrotransposition and self splicing qualities [19].

Archaeal pyruvate can produce histone deacetylase inhibition resulting in endogenous retroviral (HERV) reverse transcriptase and integrase expression. This can integrate the RNA viroidal complementary DNA into the noncoding region of eukaryotic non coding DNA using HERV integrase as has been described for borna and ebola viruses [20]. The noncoding DNA is lengthened by integrating RNA viroidal complementary DNA with the integration going on as a continuing event. The archaea genome can also get integrated into human genome using integrase as has been described for trypanosomes [21]. The integrated viroids and archaea can undergo vertical transmission and can exist as genomic parasites [20,21]. This increases the length and alters the grammar of the noncoding region producing memes or memory of acquired characters [22]. The viroidal complementary DNA can function as jumping genes producing a dynamic genome modulating DNA transcription. The RNA viroids can regulate mRNA function by RNA interference [19]. The phenomena of RNA interference can modulate T cell and B cell function, insulin signaling lipid metabolism, cell growth and differentiation, apoptosis, neuronal transmission and euchromatin/heterochromatin expression. The RNA viroids can recombine with HERV sequences and get encapsulated in microvesicles contributing to the retroviral state. Prion proteins can bind nucleic acids. The prion protein conformation is modulated by RNA viroid binding resulting in prion disease. RNA viroid induced mRNA interference can contribute to cell death in AIDS dementia, malignant transformation and autoimmunity in the acquired immunodeficiency syndrome.

The presence of muramic acid, HMG CoA reductase and cholesterol oxidase activity inhibited by antibiotics indicates the presence of bacteria with mevalonate pathway. The bacterial with mevalonate pathway include streptococcus, staphylococcus, actinomycetes, listeria, coxiella and borrelia [23] The bacteria and archaea with mevalonate pathway and cholesterol catabolism had a evolutionarily advantage and constitutes the isoprenoidal clade organism with the archaea evolving into mevalonate pathway gram positive and gram negative organism through horizontal gene transfer of viroidal and virus genes [24,25]. The isoprenoidal clade prokaryotes develop into other groups of prokaryotes via viroidal/virus as well as eukaryotic horizontal gene transfer producing bacterial speciation [26]. The RNA viroids and its complementary DNA developed into cholesterol enveloped RNA and DNA viruses like herpes, retrovirus, influenza virus, borna virus, cytomegalo virus and Ebstein Barr virus by recombining with eukaryotic and human genes resulting in viral speciation. Bacterial and viral species are ill defined and fuzzy with all of them forming one common genetic pool with frequent horizontal gene transfer and recombination. Thus the multi and unicellular eukaryote with its genes serves the purpose of prokaryotic and viral speciation. The multicellular eukaryote developed so that their endosymbiotic archaeal colonies could survive and forage better. The multicellular eukaryotes are like bacterial biofilms. The archaea and bacteria with a mevalonate pathway uses the extracellular RNA viroids and DNA viroids for quorum sensing and in the generation of symbiotic biofilm like structures which develop into multicellular eukaryotes [27,28]. The endosymbiotic archaea and bacteria with mevalonate pathway still uses the RNA viroids and DNA viroids for the regulation of muticellular eukaryote. Pollution is induced by the primitive nanoarchaea and mevalonate pathway bacteria synthesised PAH and methane leading on to redox stress. Redox stress leads to sodium potassium ATPase inhibition, inward movement of plasma membrane cholesterol, defective SREBP sensing, increased cholesterol synthesis and nanoarchaeal/mevalonate pathway bacterial growth [29]. Redox stress leads on to viroidal and archaeal multiplication. Redox stress can also lead to HERV reverse transcriptase and integrase expression. The noncoding DNA is formed of integrating RNA viroidal complementary DNA and archaea with the integration going on as a continuing event. The archaeal pox like dsDNA virus forms evolutionarily the nucleus. The integrated viroidal, archaeal and mevalonate pathway bacterial sequences can undergo vertical transmission and can exist as genomic parasites. The genomic integrated archaea, mevalonate pathway bacteria and viroids form a genomic reserve of bacteria and viruses which can recombine with human and eukaryotic genes producing bacterial and viral speciation. Bacteria and viruses have been related to the pathogenesis of acquired immunodeficiency syndrome and creutzfeldt jakob disease. Mycoplasmas have been described as co-factors in HIV infection [30]. Mycoplasma infection of the cell can result in expression of HERV sequences. Changes in the length of noncoding region especially human endogenous retroviruses and the expression of HERV sequences can contribute to the pathogenesis of AIDS syndrome [31]. The change in the length and grammar of the noncoding region produces eukaryotic speciation and individuality [32]. The integration of nanoarchaea, mevalonate pathway prokaryotes and viroids in to the eukaryotic and human genome produces a chimera which can multiply producing biofilm like multicellular structures having a mixed archaeal, viroidal, prokaryotic and eukaryotic characters which is a regression from the multicellular eukaryotic tissue This results in a new neuronal, metabolic, immune and tissue phenotype or microchimera leading to human diseases like acquired immunodeficiency syndrome and creutzfeldt jakob disease. The microchimera produces polyploidy which has been related to malignant transformation, autoimmune disease and neuronal degeneration like AIDS dementia described in acquired immunodeficiency syndrome.

Archaea and RNA viroid can bind the TLR receptor induce NFKB producing immune activation and cytokine TNF alpha secretion. The archaeal DXP and mevalonate pathway metabolites can bind γδ TCR and digoxin induced calcium signaling can activate NFKB producing chronic immune activation [2,33]. The archaea and viroid can induce chronic immune activation and generation of superantigens. Chronic immune activation can lead onto an increase in CD_4 receptor and chemokine receptor density producing a predilection to develop acquired immunodeficiency syndrome. The generation of superantigens leads to autoimmunity and increased incidence of autoimmune vasculitis and arthritis common in AIDS. The archaea and viroids can regulate the nervous system including the NMDA synaptic transmission [2]. NMDA can be activated by digoxin induced calcium oscillations, PAH and viroid induced RNA interference[2]. The cholesterol ring oxidase generated pyruvate can be converted by the GABA shunt pathway to glutamate. The archaeal cholesterol aromatase can generate serotonin [17]. Glutamatergic and serotoninergic transmission can lead to immune activation important in the pathogenesis of AIDS. NMDA excitotoxicity and neurotransmitter induced immune activation can lead onto AIDS dementia. The increased generation of serotonin and dopamine from bacterial cholesterol catabolism can lead to mood disorders and schizophreniform psychosis common in AIDS. The higher degree of integration of the archaea into the genome produces increased digoxin synthesis producing right hemispheric dominance and lesser degree producing left hemispheric dominance [2]. Right hemispheric dominance can lead to acquired immunodeficiency syndrome as has been reported previously from this laboratory. Archaea, viroids and digoxin can induce the host AKT PI3K, AMPK, HIF alpha and NFKB producing the Warburg metabolic phenotype [34]. The increased glycolytic hexokinase activity, decrease in blood ATP, leakage of cytochrome C, increase in serum pyruvate and decrease in acetyl CoA indicates the generation of the Warburg phenotype. There is induction of glycolysis, inhibition of PDH activity and mitochondrial dysfunction resulting in inefficient energetics. The increased glycolysis results in the upregulation of mitochondrial PT pore hexokinase resulting in cell proliferation and malignant transformation. The archaeal cholesterol catabolism also generates PAH which can modulate gap junction intercellular communication resulting in cell proliferation and malignant transformation. Archaeal PAH can thus induce neoplastic change. Archaeal cholesterol catabolism can deplete the cells of cholesterol leading onto polyploidy and malignant transformation. There is increased incidence of malignancies is like non-hodgkin's lymphomas and kaposi's sarcoma in AIDS. The lymphocytes depend on glycolysis for their energy needs. The increased glycolysis induced by the Warburg phenotype leads to immune activation. Lactic acid generated by increased glycolysis leads to immune stimulation. Immune stimulation is an association of AIDS syndrome. Cholesterol oxidase activity, increased glycolysis related NADPH oxidase activity and mitochondrial dysfunction generates free radicals important in the pathogenesis of AIDS. Free radicals are used by the HIV virus as messengers and increase retroviral replication and the viral load in the system. The accumulated pyruvate enters the gaba shunt pathway and is converted to citrate which is acted upon by citrate lyase and converted to acetyl CoA, used for cholesterol synthesis [34]. The pyruvate can be converted to glutamate and ammonia which is oxidised by archaea for energy needs. The increased cholesterol substrate also leads to increased archaeal growth and digoxin synthesis leading to metabolic channelling to the mevalonate pathway. Hyperdigoxinemia is important in the pathogenesis of AIDS. Digoxin can increase lymphocytic intracellular calcium which leads on to induction of NFKB and immune activation. Digoxin can also induce EGF and other growth factors resulting in oncogenesis. Digoxin can produce increased intracellular calcium related PT pore dysfunction and cell death [2]. The archaeal cholesterol catabolism generated PAH can also produce NMDA excitoxicity and cell death. The archaeal and mevalonate pathway bacteria cholesterol catabolism can deprive cholesterol from neuronal cell membrane and organelle membranes like mitochondrial, ER and lysosomal membranes producing cellular and organelle dysfunction and death. The Warburg phenotype is also important in neuronal degeneration producing AIDS dementia. The increased glycolysis results in increased generation of the enzyme glyceraldehyde 3 phosphate dehydrogenase (GAPD). GAPD can undergo polyadenylation via free radical activated PARP enzyme. The polyadenylated GAPD can undergo nuclear translocation producing nuclear cell death. All of these contribute to the genesis of neuronal degeneration and AIDS dementia. The AIDS dementia, malignant transformation, immune activation and autoimmune disease which are all part of the HIV syndrome can be related to the archaea and viroids. The cholesterol catabolism by archaea and mevalonate pathway bacteria results in cholesterol depletion from the host which has been described in AIDS. Cholesterol metabolic defects have also been described in creutzfeldt jakob disease. Thus the actinide, viroid and mevalonate pathway bacteria induced metabolic, genetic, immune and neuronal transmission changes can lead onto acquired immunodeficiency syndrome and creutzfeldt jakob disease.

References

[1] Valiathan M.S., Somers, K., Kartha, C.C. (1993). *Endomyocardial Fibrosis.* Delhi: Oxford University Press.

[2] Kurup R., Kurup, P.A. (2009). *Hypothalamic digoxin, cerebral dominance and brain function in health and diseases*. New York: Nova Science Publishers.

[3] Hanold D., Randies, J.W. (1991). Coconut cadang-cadang disease and its viroid agent, *Plant Disease,* 75, 330-335.

[4] Edwin B.T., Mohankumaran, C. (2007). Kerala wilt disease phytoplasma: Phylogenetic analysis and identification of a vector, *Proutista moesta, Physiological and Molecular Plant Pathology,* 71(1-3), 41-47.

[5] Eckburg P.B., Lepp, P.W., Relman, D.A. (2003). Archaea and their potential role in human disease, *Infect Immun,* 71, 591-596.

[6] Adam Z. (2007). Actinides and Life's Origins, *Astrobiology,* 7, 6-10.

[7] Schoner W. (2002). Endogenous cardiac glycosides, a new class of steroid hormones, *Eur. J. Biochem.,* 269, 2440-2448.

[8] Davies P.C.W., Benner, S.A., Cleland, C.E., Lineweaver, C.H., McKay, C.P., Wolfe-Simon, F. (2009). Signatures of a Shadow Biosphere, *Astrobiology,* 10, 241-249.

[9] Richmond W. (1973). Preparation and properties of a cholesterol oxidase from nocardia species and its application to the enzymatic assay of total cholesterol in serum, *Clin. Chem.,* 19, 1350-1356.

[10] Snell E.D., Snell, C.T. (1961). *Colorimetric Methods of Analysis.* Vol 3A. New York: Van NoStrand.

[11] Glick D. (1971). *Methods of Biochemical Analysis.* Vol 5. New York: Interscience Publishers.

[12] Colowick, Kaplan, N.O. (1955). *Methods in Enzymology.* Vol 2. New York: Academic Press.

[13] Maarten A.H., Marie-Jose, M., Cornelia, G., van Helden–Meewsen, Fritz, E., Marten, P.H. (1995). Detection of muramic acid in human spleen, *Infection and Immunity,* 63(5), 1652 – 1657.

[14] Smit A., Mushegian, A. (2000). Biosynthesis of isoprenoids via mevalonate in Archaea: the lost pathway, *Genome Res.,* 10(10), 1468-84.

[15] Van der Geize R., Yam, K., Heuser, T., Wilbrink, M.H., Hara, H., Anderton, M.C. (2007). A gene cluster encoding cholesterol catabolism in a soil actinomycete provides insight into Mycobacterium tuberculosis survival in macrophages, *Proc. Natl. Acad. Sci. USA,* 104(6), 1947-52.

[16] Francis A.J. (1998). Biotransformation of uranium and other actinides in radioactive wastes, *Journal of Alloys and Compounds,* 271(273), 78-84.

[17] Probian C., Wülfing, A., Harder, J. (2003). Anaerobic mineralization of quaternary carbon atoms: Isolation of denitrifying bacteria on pivalic acid (2,2-Dimethylpropionic acid), *Applied and Environmental Microbiology,* 69(3), 1866-1870.

[18] Vainshtein M., Suzina, N., Kudryashova, E., Ariskina, E. (2002). New Magnet-Sensitive Structures in Bacterial and Archaeal Cells, *Biol. Cell,* 94(1), 29-35.

[19] Tsagris E.M., de Alba, A.E., Gozmanova, M., Kalantidis, K. (2008). Viroids, *Cell Microbiol.,* 10, 2168.

[20] Horie M., Honda, T., Suzuki, Y., Kobayashi, Y., Daito, T., Oshida, T. (2010). Endogenous non-retroviral RNA virus elements in mammalian genomes, *Nature,* 463, 84-87.

[21] Hecht M., Nitz, N., Araujo, P., Sousa, A., Rosa, A., Gomes, D. (2010). Genes from Chagas parasite can transfer to humans and be passed on to children. Inheritance of DNA Transferred from American Trypanosomes to Human Hosts, *PLoS ONE,* 5, 2-10.

[22] Flam F. (1994). Hints of a language in junk DNA, *Science,* 266, 1320.

[23] Horbach S., Sahm, H., Welle, R. (1993). Isoprenoid biosynthesis in bacteria: two different pathways? *FEMS Microbiol Lett,* 111, 135–140.

[24] Gupta R.S. (1998). Protein phylogenetics and signature sequences: a reappraisal of evolutionary relationship among archaebacteria, eubacteria, and eukaryotes, *Microbiol. Mol. Biol. Rev.,* 62, 1435–1491.

[25] Margulis L. (1996). Archaeal-eubacterial mergers in the origin of Eukarya: phylogenetic classification of life. *Proc. Natl. Acad. Sci. USA,* 93, 1071-1076.

[26] Hanage W., Fraser, C., Spratt, B. (2005). Fuzzy species among recombinogenic bacteria, *BMC Biology,* 3, 6-10.

[27] Webb J.S., Givskov, M., Kjelleberg, S. (2003). Bacterial biofilms: prokaryotic adventures in multicellularity, *Curr. Opin. Microbiol.,* 6(6), 578–85.

[28] Whitchurch C.B., Tolker-Nielsen, T., Ragas, P.C., Mattick, J.S. (2002). Extracellular DNA Required for Bacterial Biofilm Formation. *Science,* 295(5559), 1487.

[29] Chen Y., Cai, T., Wang, H., Li, Z., Loreaux, E., Lingrel, J.B. (2009). Regulation of intracellular cholesterol distribution by Na/K-ATPase, *J. Biol. Chem.,* 284(22), 14881-90.

[30] Montagnier L., Blanchard, A. (1993). Mycoplasmas as cofactors in infection due to the human immunodeficiency virus. *Clin. Infect Dis.,* 17(1), S309-15.

[31] Villarreal L.P. (2006). How viruses shape the tree of life, *Future Virology,* 1(5), 587-595.

[32] Poole A.M. (2006). Did group II intron proliferation in an endosymbiont-bearing archaeon create eukaryotes? *Biol. Direct,* 1, 36-40.

[33] Eberl M., Hintz, M., Reichenberg, A., Kollas, A., Wiesner, J., Jomaa, H. (2010). Microbial isoprenoid biosynthesis and human γδ T cell activation, *FEBS Letters,* 544(1), 4-10.

[34] Wallace D.C. (2005). Mitochondria and Cancer: Warburg Addressed, *Cold Spring Harbor Symposia on Quantitative Biology,* 70, 363-374.

Chapter XIV

A Cholesterol and Actinide Dependent Archaea/Viroids in Chronic Calcific Pancreatitis and Related Disorders

Abbreviations

Muc Angio: Mucoid angiopathy
EMF: Endomyocardial fibrosis
CCP: Chronic calcific pancreatitis
MNG: Multinodular goiter

Introduction

Chronic calcific pancreatitis (CCP), endomyocardial fibrosis (EMF), multinodular goitre (MNG) and mucoid angiopathy along with the root wilt disease of coconut is endemic to Kerala with its radioactive actinide beach sands. Actinides like rutile producing intracellular magnesium deficiency due to actinide-magnesium exchange sites in the cell membrane has been implicated in the etiology of EMF [1,2,3]. Endogenous digoxin, a steroidal glycoside which functions as a membrane sodium-potassium ATPase inhibitor has also been related to its etiology of EMF, CCP, MNG and mucoid angiopathy [4]. Digoxin produces intracellular magnesium deficiency which results in acidic mucopolysaccharide accumulation of the vascular, cardiac and endocrine tissues contributing to the pathogenesis. Organisms like phytoplasmas and viroids have also been demonstrated to play a role in the etiology of root wilt disease of coconut which is co-endemic in Kerala [5,6]. The possibility of endogenous digoxin synthesis by actinide based primitive organism like archaea with a mevalonate pathway and cholesterol catabolism was considered [7,8,9]. The role of RNA viroids in the etiopathogenesis of EMF, CCP, MNG and mucoid angiopathy was also explored. Davies has put forward the concept of a shadow biosphere of organisms with alternate biochemistry present in earth itself [10]. An actinide dependent shadow biosphere of archaea and viroids in the above mentioned disease states is described [7].

Materials and Methods

Informed consent of the subjects and the approval of the ethics committee were obtained for the study. The following groups were included in the study:- endomyocardial fibrosis, chronic calcific pancreatitis, multinodular goitre and mucoid angiopathy. There were 10 patients in each group and each patient had an age and sex matched healthy control selected randomly from the general population. The blood samples were drawn in the fasting state before treatment was initiated. Plasma from fasting heparinised blood was used and the experimental protocol was as follows (I) Plasma+phosphate buffered saline, (II) same as I+cholesterol substrate, (III) same as II+rutile 0.1 mg/ml, (IV) same as II+ciprofloxacine and doxycycline each in a concentration of 1 mg/ml. Cholesterol substrate was prepared as described by Richmond [11].

Aliquots were withdrawn at zero time immediately after mixing and after incubation at 37°C for 1 hour. The following estimations were carried out:- Cytochrome F420, free RNA, free DNA, muramic acid, polycyclic aromatic hydrocarbon, hydrogen peroxide, serotonin, pyruvate, ammonia, glutamate, cytochrome C, hexokinase, ATP synthase, HMG CoA rcdutase, digoxin and urease [12,13,14,15]. Cytochrome F420 was estimated flourimetrically (excitation wavelength 420 nm and emission wavelength 520 nm). Polycyclic aromatic hydrocarbon was estimated by measuring hydrogen peroxide liberated by using glucose reagent. The statistical analysis was done by ANOVA.

Results

The parameters checked as indicated above were:- cytochrome F420, free RNA, free DNA, muramic acid, polycyclic aromatic hydrocarbon, hydrogen peroxide, serotonin, pyruvate, ammonia, glutamate, cytochrome C, hexokinase, ATP synthase, HMG CoA reductase, digoxin and urease. Plasma of control subjects showed increased levels of the above mentioned parameters with after incubation for 1 hour and addition of cholesterol substrate resulted in still further significant increase in these parameters. The plasma of patients showed similar results but the extent of increase was more. The addition of antibiotics to the control plasma caused a decrease in all the parameters while addition of rutile increased their levels. The addition of antibiotics to the patient's plasma caused a decrease in all the parameters while addition of rutile increased their levels but the extent of change was more in patient's sera as compared to controls. The results are expressed in tables 1-7 as percentage change in the parameters after 1 hour incubation as compared to the values at zero time.

Discussion

There was increase in cytochrome F420 indicating archaeal growth in endomyocardial fibrosis, chronic calcific pancreatitis, multinodular goitre and mucoid angiopathy. The archaea can synthesise and use cholesterol as a carbon and energy source [16,17]. The archaeal origin of the enzyme activities was indicated by antibiotic induced suppression. The study indicates the presence of actinide based archaea with an alternate actinide based enzymes or metalloenzymes in the system as indicated by rutile induced increase in enzyme activities [18].

There was also an increase in archaeal HMG CoA reductase activity indicating increased cholesterol synthesis by the archaeal mevalonate pathway. The archaeal beta hydroxyl steroid dehydrogenase activity indicating digoxin synthesis and archaeal cholesterol hydroxylase activity indicating bile acid synthesis were increased [8]. The archaeal cholesterol oxidase activity was increased resulting in generation of pyruvate and hydrogen peroxide [17]. The pyruvate gets converted to glutamate and ammonia by the GABA shunt pathway.

The archaeal aromatization of cholesterol generating PAH, serotonin and dopamine was also detected [19]. The archaeal glycolytic hexokinase activity and archaeal extracellular ATP synthase activity were increased. There was an increase in free RNA indicating self replicating RNA viroids and free DNA indicating generation of viroid complementary DNA strands by archaeal reverse transcriptase activity. The actinides modulate RNA folding and catalyse its ribozymal action. Digoxin can cut and paste the viroidal strands by modulating RNA splicing generating RNA viroidal diversity. The viroids are evolutionarily escaped archaeal group I introns which have retrotransposition and self splicing qualities [20]. Archaeal pyruvate can produce histone deacetylase inhibition resulting in endogenous retroviral (HERV) reverse transcriptase and integrase expression. This can integrate the RNA viroidal complementary DNA into the noncoding region of eukaryotic non coding DNA using HERV integrase as has been described for borna and ebola viruses [21]. The noncoding DNA is lengthened by integrating RNA viroidal complementary DNA with the integration going on as a continuing event. The archaea genome can also get integrated into human genome using integrase as has been described for trypanosomes [22]. The integrated viroids and archaea can undergo vertical transmission and can exist as genomic parasites [21,22]. This increases the length and alters the grammar of the noncoding region producing memes or memory of acquired characters [23]. The viroidal complementary DNA can function as jumping genes producing a dynamic genome and changing DNA sequences. The RNA viroids can regulate mrna function by RNA interference [20]. The phenomena of RNA interference can modulate euchromatin/heterochromatin expression. RNA viroidal mRNA interference plays a role in the pathogenesis of endomyocardial fibrosis, chronic calcific pancreatitis, multinodular goitre and mucoid angiopathy.

The viroidal RNA modulation of T cell and B cell function by mRNA interference can lead to immune activation. Monocytic infiltration of the vascular wall, cardiac and endocrine tissue can produce reactive connective tissue macromolecular deposition contributing to EMF, CCP, MNG and mucoid angiopathy. The viroidal RNA mediated mRNA interference can also inhibit insulin signalling and secretion leading onto CCP. The viroid RNA can inhibit thyroid hormone secretion and action by mRNA interference leading to increased TSH secretion and multinodular goitre.

The presence of muramic acid, HMG CoA reductase and cholesterol oxidase activity inhibited by antibiotics indicates the presence of bacteria with mevalonate pathway. The bacterial with mevalonate pathway include streptococcus, staphylococcus, actinomycetes, listeria, coxiella and borrelia [24].

Table 1. Effect of rutile and antibiotics on muramic acid and serotonin

	Muramic acid % (Increase without Doxy)		Muramic acid % (Decrease with Doxy)		5 HT % (Increase without Doxy)		5 HT % (Decrease with Doxy)	
Group	Mean	± SD	Mean	± SD	Mean	± SD	Mean	± SD
Normal	4.41	0.15	18.63	0.12	4.34	0.15	18.24	0.37
Muc Angio	24.43	0.81	68.72	2.77	24.32	1.09	65.80	5.14
EMF	22.28	1.52	64.05	2.79	22.82	1.56	64.61	4.95
CCP	23.07	1.46	64.68	3.86	22.89	1.50	64.19	6.51
MNG	23.85	1.69	66.43	3.17	22.72	1.64	63.91	4.93
F value	403.394		680.284		348.867		364.999	
P value	< 0.001		< 0.001		< 0.001		< 0.001	

Table 2. Effect of rutile and antibiotics on free DNA and RNA

	DNA % change (Increase with Rutile)		DNA % change (Decrease with Doxy)		RNA % change (Increase with Rutile)		RNA % change (Decrease with Doxy)	
Group	Mean	± SD	Mean	± SD	Mean	± SD	Mean	± SD
Normal	4.37	0.15	18.39	0.38	4.37	0.13	18.38	0.48
Muc Angio	22.27	1.49	63.99	4.03	22.27	1.49	69.25	2.33
EMF	22.29	2.05	58.70	7.34	22.29	2.05	67.03	5.97
CCP	21.19	2.18	61.63	7.68	21.19	2.18	62.99	5.47
MNG	22.93	2.08	63.49	5.01	23.19	1.74	65.68	4.06
F value	337.577		356.621		427.828		654.453	
P value	< 0.001		< 0.001		< 0.001		< 0.001	

Table 3. Effect of rutile and antibiotics on HMG CoA reductase and PAH

	HMG CoA R % change (Increase with Rutile)		HMG CoA R % change (Decrease with Doxy)		PAH % change (Increase with Rutile)		PAH % change (Decrease with Doxy)	
Group	Mean	± SD	Mean	± SD	Mean	± SD	Mean	± SD
Normal	4.30	0.20	18.35	0.35	4.45	0.14	18.25	0.72
Muc Angio	24.44	0.90	59.90	4.74	23.90	1.36	63.29	6.86
EMF	22.92	1.48	61.91	7.56	23.73	1.38	65.20	6.20
CCP	23.27	1.96	63.09	9.21	22.85	1.71	66.14	3.58

	HMG CoA R % change (Increase with Rutile)		HMG CoA R % change (Decrease with Doxy)		PAH % change (Increase with Rutile)		PAH % change (Decrease with Doxy)	
Group	Mean	± SD	Mean	± SD	Mean	± SD	Mean	± SD
MNG	23.65	1.88	64.78	6.62	23.79	1.19	64.24	3.96
F value	319.332		199.553		391.318		257.996	
P value	< 0.001		< 0.001		< 0.001		< 0.001	

Table 4. Effect of rutile and antibiotics on digoxin and urease

	Digoxin (ng/ml) (Increase with Rutile)		Digoxin (ng/ml) (Decrease with Doxy+Cipro)		Urease % change (Increase with Rutile)		Urease % change (Decrease with Doxy)	
Group	Mean	± SD	Mean	± SD	Mean	± SD	Mean	± SD
Normal	0.11	0.00	0.054	0.003	4.29	0.18	18.15	0.58
Muc Angio	0.53	0.03	0.224	0.041	23.37	1.55	63.99	4.03
EMF	0.51	0.05	0.213	0.033	23.41	1.41	58.70	7.34
CCP	0.47	0.05	0.212	0.028	22.44	2.00	61.63	7.68
MNG	0.51	0.06	0.227	0.040	22.15	1.79	65.49	7.28
F value	135.116		71.706		290.441		203.651	
P value	< 0.001		< 0.001		< 0.001		< 0.001	

Table 5. Effect of rutile and antibiotics on pyruvate and hexokinase

	Pyruvate % change (Increase with Rutile)		Pyruvate % change (Decrease with Doxy)		Hexokinase % change (Increase with Rutile)		Hexokinase % change (Decrease with Doxy)	
Group	Mean	± SD	Mean	± SD	Mean	± SD	Mean	± SD
Normal	4.34	0.21	18.43	0.82	4.21	0.16	18.56	0.76
Muc Angio	22.27	1.49	61.94	5.49	23.67	1.65	69.25	2.33
EMF	22.29	2.05	62.37	5.05	21.66	1.94	67.03	5.97
CCP	21.19	2.18	54.82	8.70	22.27	2.18	62.99	5.47
MNG	19.73	2.27	59.36	7.53	22.51	2.32	62.70	3.24
F value	321.255		115.242		292.065		317.966	
P value	< 0.001		< 0.001		< 0.001		< 0.001	

Table 6. Effect of rutile and antibiotics on hydrogen peroxide and delta amino levulinic acid

	H_2O_2 % (Increase with Rutile)		H_2O_2 % (Decrease with Doxy)		ALA % (Increase with Rutile)		ALA % (Decrease with Doxy)	
Group	Mean	± SD	Mean	± SD	Mean	± SD	Mean	± SD
Normal	4.43	0.19	18.13	0.63	4.40	0.10	18.48	0.39
Muc Angio	23.64	1.50	60.44	6.83	22.27	1.49	59.90	4.74
EMF	23.29	1.67	60.52	5.38	22.29	2.05	61.91	7.56
CCP	23.38	1.79	57.37	7.45	21.19	2.18	63.09	9.21
MNG	22.00	1.77	61.39	7.47	22.71	1.82	66.13	3.83
F value	380.721		171.228		372.716		556.411	
P value	< 0.001		< 0.001		< 0.001		< 0.001	

Table 7. Effect of rutile and antibiotics on ATP synthase and cytochrome F 420

	ATP synthase % (Increase with Rutile)		ATP synthase % (Decrease with Doxy)		CYT F420 % (Increase with Rutile)		CYT F420 % (Decrease with Doxy)	
Group	Mean	± SD	Mean	± SD	Mean	± SD	Mean	± SD
Normal	4.40	0.11	18.78	0.11	4.48	0.15	18.24	0.66
Muc Angio	23.45	1.52	67.05	4.84	23.72	1.76	58.92	5.46
EMF	23.37	1.31	63.97	3.62	22.70	1.87	60.46	8.06
CCP	22.53	1.92	66.31	3.10	21.31	1.37	57.32	8.41
MNG	23.39	1.14	68.11	3.02	22.17	2.01	65.15	6.46
F value	449.503		673.081		306.749		130.054	
P value	< 0.001		< 0.001		< 0.001		< 0.001	

The bacteria and archaea with mevalonate pathway and cholesterol catabolism had a evolutionarily advantage and constitutes the isoprenoidal clade organism with the archaea evolving into mevalonate pathway gram positive and gram negative organism through horizontal gene transfer of viroidal and virus genes [25]. The isoprenoidal clade prokaryotes develop into other groups of prokaryotes via viroidal/virus as well as eukaryotic horizontal gene transfer producing bacterial speciation [26]. The RNA viroids and its complementary DNA developed into cholesterol enveloped RNA and DNA viruses like herpes, retrovirus, influenza virus, borna virus, cytomegalo virus and Ebstein Barr virus by recombining with eukaryotic and human genes resulting in viral speciation. Bacterial and viral species are ill defined and fuzzy with all of them forming one common genetic pool with frequent horizontal gene transfer and recombination. Thus the multi and unicellular eukaryote with its genes serves the purpose of prokaryotic and viral speciation. The multicellular eukaryote developed so that their endosymbiotic archaeal colonies could survive and forage better. The multicellular eukaryotes are like bacterial biofilms. The archaea and bacteria with a mevalonate pathway uses the extracellular RNA viroids and DNA viroids for quorum sensing and in the generation of symbiotic biofilm like structures which develop into multicellular eukaryotes [27,28]. The endosymbiotic archaea and bacteria with mevalonate pathway still uses the RNA viroids and DNA viroids for the regulation of muticellular eukaryote. Pollution is induced by the primitive nanoarchaea and mevalonate pathway bacteria synthesised PAH and methane leading on to redox stress. Redox stress leads to sodium potassium ATPase inhibition, inward movement of plasma membrane cholesterol, defective SREBP sensing, increased cholesterol synthesis and nanoarchaeal/mevalonate pathway bacterial growth [29]. Redox stress leads on to viroidal and archaeal multiplication. Redox stress can also lead to HERV reverse transcriptase and integrase expression. The noncoding DNA is formed of integrating RNA viroidal complementary DNA and archaea with the integration going on as a continuing event. The change in the length and grammar of the noncoding region produces eukaryotic speciation and individuality [30]. Changes in the length of noncoding region especially due to integration of viroid complementary DNA and archaea and the resulting jumping genes leads to new DNA sequences possibly contributing to EMF, CCP, MNG and mucoid angiopathy [31]. The integrated viroidal, archaeal and mevalonate pathway bacterial sequences can undergo vertical transmission and can exist as genomic parasites. The genomic integrated archaea, mevalonate pathway bacteria and viroids form a genomic reserve of bacteria and viruses which can recombine with human and eukaryotic genes producing bacterial and viral speciation. Archaea and mevalonate pathway bacteria can lead onto EMF, CCP, MNG and mucoid angiopathy. The persistent symbiosis leads to reparative connective tissue macromolecular deposition of acidic mucopolysaccharides, glycoproteins, collagen and elastin leading to fibrotic changes in the heart, vessel wall, thyroid and pancreas contributing to EMF, CCP, MNG and mucoid angiopathy [4,32]. The integration of nanoarchaea, mevalonate pathway prokaryotes and viroids in to the eukaryotic and human genome produces a chimera which can multiply producing biofilm like multicellular structures having a mixed archaeal, viroidal, prokaryotic and eukaryotic characters which is a regression from the multicellular eukaryotic tissue. This results in a new metabolic and immune phenotype or microchimeras leading on to human diseases like EMF, CCP, MNG and mucoid angiopathy with a predilection to develop malignancy. Microchimeras can lead to cellular polyploidy important in malignant transformation and induction of carcinoma of thyroid and pancreas. The growth of archaea in the vascular, cardiac and endocrine tissues can result in calcification. The archaea can form calcified nanoarchaeal structures which can exist as colonies in slime. The archaea can undergo magnetite and calcium carbonate mineralization and can exist as calcified nanoforms [33]. The calcified nanoarchaea can contribute to the tissue calcification noted in CCP, MNG and mucoid angiopathy.

Archaea and RNA viroid can bind the TLR receptor induce NFKB producing immune activation and cytokine TNF alpha secretion. The archaeal DXP and mevalonate pathway metabolites can bind γδ TCR and digoxin induced calcium signaling can activate NFKB producing chronic immune activation [4,34]. The archaea and viroid can induce chronic immune activation and generation of superantigens. The archaea and viroid induced chronic immune activation can lead to monocyte infiltration of the vessel wall, cardiac and endocrine tissues leading on to reparative connective tissue macromolecular deposition. Immune activation results in induction of NADPH oxidase which generates hydrogen peroxide. Cholesterol oxidase activity also generates hydrogen peroxide. Hydrogen peroxide can produce tissue injury in MNG, CCP, EMF and mucoid angiopathy contributing to reparative connective tissue macromolecular deposition. Immune activation can also produce insulin resistance. TNF alpha produced by chronic immune activation can modulate the insulin receptor producing insulin resistance [35]. Chronic immune activation and cholesterol oxidase generated hydrogen peroxide can induce neutral sphingomyelinase generating ceramide producing insulin resistance [36]. This can contribute to chronic calcific pancreatitis. Immune activation and NFKB induction can suppress the thyroid hormone receptor resulting in hypothyroidism and increased TSH levels contributing to thyroid gland enlargement and multinodular goitre. Immune activation and NFKB induction can suppress the nuclear receptors LXR, PXR and FXR. FXR suppression can also lead to insulin resistance as well as increased connective tissue MPS deposition in vessel wall, cardiac tissue and endocrine tissue. LXR suppression by NFKB stimulates HMG CoA reductase activity and suppresses cholesterol 7 alpha hydroxylase activity [37]. This stimulates cholesterol synthesis and inhibits its degradation via the bile acid pathway. PXR suppression by NFKB prevents cholesterol detoxification via the bile acid shunt pathway [38]. Thus LXR and PXR suppression by NFKB produces acute cholesterol toxicity. The increased cholesterol in the system leads to still further archaeal multiplication and growth as they depend on cholesterol as a carbon and energy source.

Archaea, viroids and digoxin can induce the host AKT PI3K, AMPK, HIF alpha and NFKB producing the Warburg metabolic phenotype [39]. The increased glycolytic hexokinase activity, decrease in blood ATP, leakage of cytochrome C, increase in serum pyruvate and decrease in acetyl CoA indicates the generation of the Warburg phenotype. There is induction of glycolysis, inhibition of PDH activity and mitochondrial dysfunction resulting in inefficient energetics. Mitochondrial dysfunction owing to the Warburg's phenotype can contribute to ineffective glucose utilisation and CCP. The accumulated pyruvate enters the gaba shunt pathway and is converted to citrate which is acted upon by citrate lyase and converted to acetyl CoA, used for cholesterol synthesis [39]. The increased cholesterol substrate also leads to increased archaeal growth and digoxin synthesis due to metabolic channeling to the mevalonate pathway. The Warburg phenotype leads to increased lipid synthesis and defective beta oxidation of fatty acids. The myocardium depends on fatty acids beta oxidation for energetics. The defective beta oxidation of fatty acids leads to myocardial dysfunction and EMF. The Warburg phenotype leads to upregulated glycolysis and increase in the metabolite fructose 1,6 diphosphate which is channelled to the pentose phosphate pathway. This can generate UDP sugars used for mucopolysaccharide synthesis. This results in acidic MPS deposition in the tissues leading onto EMF, CCP, MNG and mucoid angiopathy. The pyruvate can be converted to glutamate and ammonia which is oxidised by archaea for energy needs. Ammonia can stimulate membrane sodium-potassium ATPase, increase ATP utilisation and produce mitochondrial transmembrane potential changes leading to mitochondrial dysfunction. This causes defective glucose utilisation contributing to CCP. Archaeal urease can convert urea to ammonia and thiocyanate. Increase cyanide load in the system can lead to mitochondrial dysfunction [3]. Cyanide related mitochondrial dysfunction can produce EMF, CCP and MNG. It produces defective cardiac function, decreased glucose utilisation and impaired iodide transport into the thyroid follicular cells. The Warburg phenotype can also lead onto malignant transformation. The upregulated glycolysis results in increased mitochondrial PT pore hexokinase and cell proliferation producing carcinoma of thyroid and pancreas.

Digoxin can produce sodium-potassium ATPase inhibition and inward movement of plasma membrane cholesterol. This produces defective SREBP sensing, increased HMG CoA reductase activity and cholesterol synthesis [29]. The digoxin induced inward movement of plasma membrane cholesterol can alter membrane cholesterol/sphingomyelin ratio producing modified lipid microdomains [40]. The digoxin induced lipid microdomain modulation can regulate the GPCR couple adrenaline, noradrenaline, glucagon and neuropeptide receptors as well as protein tyrosine kinase linked insulin receptor. This can lead onto CCP. The digoxin mediated inhibition of nuclear membrane sodium-potassium ATPase can modulate nuclear membrane lipid microdomains and thyroxine DNA receptor function. This can lead onto hypothyroidism, increased TSH levels and thyroid gland enlargement contributing to MNG. Digoxin can produce intracellular hypercalcemia and hypomagnesemia. This can lead on to vasospasm and thrombosis. Intracellular hypercalcemia can activate the G-protein coupled thrombin receptor and PAF receptor producing thrombosis. Intracellular magnesium deficiency can lead onto increased thrombin and ADP/collagen induced platelet aggregation. This leads onto the thrombotic state in mucoid angiopathy. The decreased intracellular magnesium can produce ATP synthase inhibition and the increased intracellular calcium can produce mitochondrial PT pore dysfunction. Mitochondrial dysfunction can contribute to decreased glucose utilisation in CCP and myocardial dysfunction in EMF. Digoxin can produce sodium-potassium ATPase inhibition and intracellular hypomagnesemia. The increased tissue rutile load can lead to rutile-magnesium exchange leading onto intracellular hypomagnesemia. Hypomagnesemia can lead onto upregulated connective tissue macromolecular synthesis contributing to MNG, CCP, EMF and mucoid angiopathy. Acidic MPS deposition in the vessel wall leads to a hose pipe narrowing of the entire vascular tree leading onto mucoid angiopathy. Acidic MPS, collagen and elastin deposition of the heart leads to EMF. Hyperdigoxinemia is important in the pathogenesis of EMF, CCP, MNG and mucoid angiopathy. Digoxin induced sodium-potassium ATPase inhibition results in an ATP sparing effect [41]. Eighty percent of the ATP generated is used to run the sodium-potassium ATPase pump. The digoxin inhibition of the sodium-potassium ATPase spares this ATP which is then used for lipid and cholesterol synthesis. Fat also fuels insulin resistance by binding to the toll receptor and producing immune activation and immune infiltration of the adipose tissue. Digoxin can also increase lymphocytic intracellular calcium which leads on to induction of NFKB and immune activation [4]. The archaeal cholesterol catabolism can deplete the lymphocytic cell membranes of cholesterol resulting in alteration of lymphocytic cell membrane microdomains related receptors producing immune activation, monocytic infiltration and reparative connective tissue macromolecular deposition.

NMDA can be activated by digoxin induced calcium oscillations, PAH and viroid induced RNA interference [4]. The cholesterol ring oxidase generated pyruvate can be converted by the GABA shunt pathway to glutamate. Glutamatergic transmission can lead to immune activation. Immune activation can lead to reparative connective tissue macromolecular deposition in EMF, CCP, MNG and mucoid angiopathy. The cholesterol aromatase generated serotonin is well known to produce connective tissue macromolecule especially collagen deposition producing the fibrotic changes in EMF, mucoid angiopathy, MNG and CCP. The archaeal cholesterol aromatase can generate PAH [19]. The PAH can also lead to insulin resistance and CCP. PAH can also inhibit thyroid hormone receptor function contributing to hypothyroidism, increased TSH, thyroid enlargement and MNG. Particulate pollution has been related to vascular thrombosis and can lead to mucoid angiopathy. PAH particles are also known to produce myocardial dysfunction. Thus the actinide, viroid and mevalonate pathway bacteria induced metabolic, genetic, immune and neuronal transmission changes can lead onto endemic EMF, CCP, MNG and mucoid angiopathy. The term archaea and viroid induced endemic cardiovascular and endocrine mucopolysaccharidoses can be used to describe this entity.

References

[1] Sandhyamoni S. (1993). Mucoid vasculopathy- vascular lesions in autopsy studies, *Mod. Pathology,* 6, 341-349.

[2] Balakrishnan V. (1987). *Chronic Pancreatitis in India*. Delhi: Indian Society of Pancreatology.

[3] Valiathan M.S., Somers, K., Kartha, C.C. (1993). *Endomyocardial Fibrosis.* Delhi: Oxford University Press.

[4] Kurup R., Kurup, P.A. (2009). *Hypothalamic digoxin, cerebral dominance and brain function in health and diseases*. New York: Nova Science Publishers.

[5] Hanold D., Randies, J.W. (1991). Coconut cadang-cadang disease and its viroid agent, *Plant Disease,* 75, 330-335.

[6] Edwin B.T., Mohankumaran, C. (2007). Kerala wilt disease phytoplasma: Phylogenetic analysis and identification of a vector, *Proutista moesta, Physiological and Molecular Plant Pathology,* 71(1-3), 41-47.

[7] Adam Z. (2007). Actinides and Life's Origins, *Astrobiology,* 7, 6-10.

[8] Schoner W. (2002). Endogenous cardiac glycosides, a new class of steroid hormones, *Eur. J. Biochem.,* 269, 2440-2448.

[9] Eckburg P.B., Lepp, P.W., Relman, D.A. (2003). Archaea and their potential role in human disease, *Infect. Immun.,* 71, 591-596.

[10] Davies P.C.W., Benner, S.A., Cleland, C.E., Lineweaver, C.H., McKay, C.P., Wolfe-Simon, F. (2009). Signatures of a Shadow Biosphere, *Astrobiology,* 10, 241-249.

[11] Richmond W. (1973). Preparation and properties of a cholesterol oxidase from nocardia species and its application to the enzymatic assay of total cholesterol in serum, *Clin. Chem.,* 19, 1350-1356.

[12] Snell E.D., Snell, C.T. (1961). *Colorimetric Methods of Analysis.* Vol 3A. New York: Van NoStrand.

[13] Glick D. (1971). *Methods of Biochemical Analysis.* Vol 5. New York: Interscience Publishers.

[14] Colowick, Kaplan, N.O. (1955). *Methods in Enzymology.* Vol 2. New York: Academic Press.

[15] Maarten A.H., Marie-Jose, M., Cornelia, G., van Helden–Meewsen, Fritz, E., Marten, P.H. (1995). Detection of muramic acid in human spleen, *Infection and Immunity,* 63(5), 1652 – 1657.

[16] Smit A., Mushegian, A. (2000). Biosynthesis of isoprenoids via mevalonate in Archaea: the lost pathway, *Genome Res.,* 10(10), 1468-84.

[17] Van der Geize R., Yam, K., Heuser, T., Wilbrink, M.H., Hara, H., Anderton, M.C. (2007). A gene cluster encoding cholesterol catabolism in a soil actinomycete provides insight into Mycobacterium tuberculosis survival in macrophages, *Proc. Natl. Acad. Sci. USA,* 104(6), 1947-52.

[18] Francis A.J. (1998). Biotransformation of uranium and other actinides in radioactive wastes, *Journal of Alloys and Compounds,* 271(273), 78-84.

[19] Probian C., Wülfing, A., Harder, J. (2003). Anaerobic mineralization of quaternary carbon atoms: Isolation of denitrifying bacteria on pivalic acid (2,2-Dimethylpropionic acid), *Applied and Environmental Microbiology,* 69(3), 1866-1870.

[20] Tsagris E.M., de Alba, A.E., Gozmanova, M., Kalantidis, K. (2008). Viroids, *Cell Microbiol.,* 10, 2168.

[21] Horie M., Honda, T., Suzuki, Y., Kobayashi, Y., Daito, T., Oshida, T. (2010). Endogenous non-retroviral RNA virus elements in mammalian genomes, *Nature,* 463, 84-87.

[22] Hecht M., Nitz, N., Araujo, P., Sousa, A., Rosa, A., Gomes, D. (2010). Genes from Chagas parasite can transfer to humans and be passed on to children. Inheritance of DNA Transferred from American Trypanosomes to Human Hosts, *PLoS ONE,* 5, 2-10.

[23] Flam F. (1994). Hints of a language in junk DNA, *Science,* 266, 1320.

[24] Horbach S., Sahm, H., Welle, R. (1993). Isoprenoid biosynthesis in bacteria: two different pathways? *FEMS Microbiol. Lett,* 111, 135–140.

[25] Gupta R.S. (1998). Protein phylogenetics and signature sequences: a reappraisal of evolutionary relationship among archaebacteria, eubacteria, and eukaryotes, *Microbiol. Mol. Biol. Rev.,* 62, 1435–1491.

[26] Hanage W., Fraser, C., Spratt, B. (2005). Fuzzy species among recombinogenic bacteria, *BMC Biology,* 3, 6-10.

[27] Webb J.S., Givskov, M., Kjelleberg, S. (2003). Bacterial biofilms: prokaryotic adventures in multicellularity, *Curr. Opin. Microbiol.,* 6(6), 578–85.

[28] Whitchurch C.B., Tolker-Nielsen, T., Ragas, P.C., Mattick, J.S. (2002). Extracellular DNA Required for Bacterial Biofilm Formation. *Science,* 295(5559), 1487.

[29] Chen Y., Cai, T., Wang, H., Li, Z., Loreaux, E., Lingrel, J.B. (2009). Regulation of intracellular cholesterol distribution by Na/K-ATPase, *J. Biol. Chem.,* 284(22), 14881-90.

[30] Poole A.M. (2006). Did group II intron proliferation in an endosymbiont-bearing archaeon create eukaryotes? *Biol. Direct,* 1, 36-40.

[31] Villarreal L.P. (2006). How viruses shape the tree of life, *Future Virology,* 1(5), 587-595.

[32] Khovidhunkit W., Kim, M.S., Memon, R.A., Shigenaga, J.K., Moser, A.H., Feingold, K.R. (2004). Thematic review series: The Pathogenesis of Atherosclerosis. Effects of infection and inflammation on lipid and lipoprotein metabolism mechanisms and consequences to the host, *J Lipid Res,* 45(7), 1169 - 1196.

[33] Vainshtein M., Suzina, N., Kudryashova, E., Ariskina, E. (2002). New Magnet-Sensitive Structures in Bacterial and Archaeal Cells, *Biol. Cell,* 94(1), 29-35.

[34] Eberl M., Hintz, M., Reichenberg, A., Kollas, A., Wiesner, J., Jomaa, H. (2010). Microbial isoprenoid biosynthesis and human γδ T cell activation, *FEBS Letters,* 544(1), 4-10.

[35] Cani P.D., Amar, J., Iglesias, M.A., Poggi, M., Knauf, C., Bastelica, D. (2007). Metabolic endotoxemia initiates obesity and insulin resistance, *Diabetes,* 56,1761–1772.

[36] Memon R.A., Holleran, W.M., Moser, A.H., Seki, T., Uchida, Y., Fuller, J. (1998). Endotoxin and Cytokines Increase Hepatic Sphingolipid Biosynthesis and Produce Lipoproteins Enriched in Ceramides and Sphingomyclin, *Arterioscler Thromb. Vasc. Biol.,* 18(8), 1257 - 1265.

[37] Carayol N., Chen, J., Yang, F., Jin, T., Jin, L., States, D. (2006). A dominant function of IKK/NF-kB signaling in global LPS-induced gene expression, *J. Biol. Chem.,* 10, 1074.

[38] Kliewer S.A. (2005). Cholesterol detoxification by the nuclear pregnane X receptor, *Proc. Natl. Acad. Sci. USA,* 102(8), 2675-6.

[39] Wallace D.C. (2005). Mitochondria and Cancer: Warburg Addressed, *Cold Spring Harbor Symposia on Quantitative Biology,* 70, 363-374.

[40] Paila Y.D., Tiwari, S., Chattopadhyay, A. (2009). Are specific nonannular cholesterol binding sites present in G-protein coupled receptors? *Biochim. Biophys. Acta,* 1788(2), 295-302.

[41] Ebensperger G., Ebensperger, R., Herrera, E.A., Riquelme, R.A., Sanhueza, E.M., Lesage, F. (2005). Fetal brain hypometabolism during prolonged hypoxaemia in the llama, *J. Physiol.*, 567(3), 963 – 975.

Chapter XV

A Cholesterol and Actinide Dependent Archaea and Viroids in Gastrointestinal and Hepatic Diseases

Abbreviations

PUD: Peptic ulcer disease
UC: Ulcerative colitis
IBS: Irritable bowel syndrome

Introduction

Endomyocardial fibrosis (EMF) along with the root wilt disease of coconut is endemic to Kerala with its radioactive actinide beach sands. Actinides like rutile producing intracellular magnesium deficiency due to rutile-magnesium exchange sites in the cell membrane has been implicated in the etiology of EMF [1]. Endogenous digoxin, a steroidal glycoside which functions as a membrane sodium-potassium ATPase inhibitor has also been related to its etiology due to the intracellular magnesium deficiency it produces [2]. Organisms like phytoplasmas and viroids have also been demonstrated to play a role in the etiology of these diseases [3,4]. Endogenous digoxin has been related to the pathogenesis of cirrhosis liver, ulcerative colitis, irritable bowel syndrome and peptic ulcer disease [2]. The possibility of endogenous digoxin synthesis by actinide based primitive organism like archaea with a mevalonate pathway and cholesterol catabolism was considered [5,6,7]. Davies has put forward the concept of a shadow biosphere of organisms with alternate biochemistry present in earth itself [8]. An actinide dependent shadow biosphere of archaea and viroids in the above mentioned disease states is described [6].

Materials and Methods

Informed consent of the subjects and the approval of the ethics committee were obtained for the study. The following groups were included in the study:- cirrhosis liver, ulcerative colitis, irritable bowel syndrome and peptic ulcer disease. There were 10 patients in each group and each patient had an age and sex matched healthy control selected randomly from the general population. The blood samples were drawn in the fasting state before treatment was initiated. Plasma from fasting heparinised blood was used and the experimental protocol was as follows (I) Plasma+phosphate buffered saline, (II) same as I+cholesterol substrate, (III) same as II+rutile 0.1 mg/ml, (IV) same as II+ciprofloxacine and doxycycline each in a concentration of 1 mg/ml. Cholesterol substrate was prepared as described by Richmond [9]. Aliquots were withdrawn at zero time immediately after mixing and after incubation at 37°C for 1 hour. The following estimations were carried out:- Cytochrome F420, free RNA, free DNA, muramic acid, polycyclic aromatic hydrocarbon, hydrogen peroxide, serotonin, pyruvate, ammonia, glutamate, cytochrome C, hexokinase, ATP synthase, HMG CoA redutase, digoxin and bile acids [10,11,12,13]. Cytochrome F420 was estimated flourimetrically (excitation wavelength 420 nm and emission wavelength 520 nm). Polycyclic aromatic hydrocarbon was estimated by measuring hydrogen peroxide liberated by using glucose reagent. The statistical analysis was done by ANOVA.

Results

The parameters checked as indicated above were:- cytochrome F420, free RNA, free DNA, muramic acid, polycyclic aromatic hydrocarbon, hydrogen peroxide, serotonin, pyruvate, ammonia, glutamate, cytochrome C, hexokinase, ATP synthase, HMG CoA redutase, digoxin and bile acids. Plasma of control subjects showed increased levels of the above mentioned parameters with after incubation for 1 hour and addition of cholesterol substrate resulted in still further significant increase in these parameters. The plasma of patients showed similar results but the extent of increase was more. The addition of antibiotics to the control plasma caused a decrease in all the parameters while addition of rutile increased their levels. The addition of antibiotics to the patient's plasma caused a decrease in all the parameters while addition of rutile increased their levels but the extent of change was more in patient's sera as compared to controls. The results are expressed in tables 1-7 as percentage change in the parameters after 1 hour incubation as compared to the values at zero time.

Discussion

There was increase in cytochrome F420 indicating archaeal growth in cirrhosis liver, ulcerative colitis, irritable bowel syndrome and peptic ulcer disease. The archaea can synthesise and use cholesterol as a carbon and energy source [14,15]. The archaeal origin of the enzyme activities was indicated by antibiotic induced suppression. The study indicates the presence of actinide based archaea with an alternate actinide based enzymes or metalloenzymes in the system as indicated by rutile induced increase in enzyme activities [16]. There was also an increase in archaeal HMG CoA reductase activity indicating increased cholesterol synthesis by the archaeal mevalonate pathway. The archaeal beta hydroxyl steroid dehydrogenase activity indicating digoxin synthesis and archaeal cholesterol hydroxylase activity indicating bile acid synthesis were increased [7]. The archaeal cholesterol oxidase activity was increased resulting in generation of pyruvate and hydrogen peroxide [15]. The pyruvate gets converted to glutamate and ammonia by the GABA shunt pathway.

The archaeal aromatization of cholesterol generating PAH, serotonin and dopamine was also detected [17]. The archaeal glycolytic hexokinase activity and archaeal extracellular ATP synthase activity were increased. The archaea can undergo magnetite and calcium carbonate mineralization and can exist as calcified nanoforms [18].

There was an increase in free RNA indicating self replicating RNA viroids and free DNA indicating generation of viroid complementary DNA strands by archaeal reverse transcriptase activity.

The actinides modulate RNA folding and catalyse its ribozymal action. Digoxin can cut and paste the viroidal strands by modulating RNA splicing generating RNA viroidal diversity. The viroids are evolutionarily escaped archaeal group I introns which have retrotransposition and self splicing qualities [19]. Archaeal pyruvate can produce histone deacetylase inhibition resulting in endogenous retroviral (HERV) reverse transcriptase and integrase expression. This can integrate the RNA viroidal complementary DNA into the noncoding region of eukaryotic non coding DNA using HERV integrase as has been described for borna and ebola viruses [20]. The noncoding DNA is lengthened by integrating RNA viroidal complementary DNA with the integration going on as a continuing event. The archaea genome can also get integrated into human genome using integrase as has been described for trypanosomes [21]. The integrated viroids and archaea can undergo vertical transmission and can exist as genomic parasites [20,21]. This increases the length and alters the grammar of the noncoding region producing memes or memory of acquired characters [22]. The viroidal complementary DNA can function as jumping genes producing a dynamic genome important in HLA gene expression. This modulation of HLA gene expression by viroidal complementary DNA can result in immune activation. The RNA viroids can regulate mrna function by RNA interference [19]. The phenomena of RNA interference can modulate T cell and B cell function and euchromatin/heterochromatin expression. RNA viroidal mRNA interference related immune activation plays a role in the pathogenesis of cirrhosis liver, ulcerative colitis, irritable bowel syndrome and peptic ulcer disease.

The presence of muramic acid, HMG CoA reductase and cholesterol oxidase activity inhibited by antibiotics indicates the presence of bacteria with mevalonate pathway. The bacterial with mevalonate pathway include streptococcus, staphylococcus, actinomycetes, listeria, coxiella and borrelia [23]. The bacteria and archaea with mevalonate pathway and cholesterol catabolism had a evolutionarily advantage and constitutes the isoprenoidal clade organism with the archaea evolving into mevalonate pathway gram positive and gram negative organism through horizontal gene transfer of viroidal and virus genes [24]. The isoprenoidal clade prokaryotes develop into other groups of prokaryotes via viroidal/virus as well as eukaryotic horizontal gene transfer producing bacterial speciation [25].

Table 1. Effect of rutile and antibiotics on muramic acid and serotonin

	Muramic acid % change (Increase with Rutile)		Muramic acid % change (Decrease with Doxy+Cipro)		5 HT % (Increase without Doxy)		5 HT % (Decrease with Doxy)	
Group	Mean	± SD	Mean	± SD	Mean	± SD	Mean	± SD
Normal	4.41	0.15	18.63	0.12	4.34	0.15	18.24	0.37
Cirrhosis	23.11	1.82	66.96	3.79	23.13	1.78	64.88	4.96
PUD	23.43	1.59	65.71	4.01	22.92	1.71	65.58	4.74
UC	23.81	1.45	66.85	3.72	22.83	1.96	63.42	5.10
IBS	23.28	1.95	66.02	3.90	22.79	1.79	62.70	5.05
F value	403.394		680.284		348.867		364.999	
P value	< 0.001		< 0.001		< 0.001		< 0.001	

Table 2. Effect of rutile and antibiotics on free DNA and RNA

	DNA % change (Increase with Rutile)		DNA % change (Decrease with Doxy)		RNA % change (Increase with Rutile)		RNA % change (Decrease with Doxy)	
Normal	4.37	0.15	18.39	0.38	4.37	0.13	18.38	0.48
Cirrhosis	22.78	1.94	63.06	6.20	22.91	1.69	66.23	3.44
PUD	23.07	1.50	62.99	5.27	23.32	1.92	66.07	4.11
UC	23.28	1.93	61.81	2.75	22.89	1.85	66.33	3.73
IBS	23.61	1.53	67.77	3.23	22.94	1.88	65.84	4.20
F value	337.577		356.621		427.828		654.453	
P value	< 0.001		< 0.001		< 0.001		< 0.001	

Table 3. Effect of rutile and antibiotics on HMG CoA reductase and PAH

	HMG CoA R % change (Increase with Rutile)		HMG CoA R % change (Decrease with Doxy)		PAH % change (Increase with Rutile)		PAH % change (Decrease with Doxy)	
Normal	4.30	0.20	18.35	0.35	4.45	0.14	18.25	0.72
Cirrhosis	23.29	1.67	59.19	7.18	23.39	1.63	65.88	5.01
PUD	23.56	1.83	63.61	6.60	23.06	1.56	64.49	4.64
UC	23.24	1.79	63.55	8.01	23.49	1.48	64.96	5.02
IBS	23.66	1.47	66.11	6.52	23.32	1.46	62.95	7.18
F value	319.332		199.553		391.318		257.996	
P value	< 0.001		< 0.001		< 0.001		< 0.001	

Table 4. Effect of rutile and antibiotics on digoxin and bile acids

	Digoxin (ng/ml) (Increase with Rutile)		Digoxin (ng/ml) (Decrease with Doxy+Cipro)		Bile acids % change (Increase with Rutile)		Bile acids % change (Decrease with Doxy)	
Normal	0.11	0.00	0.054	0.003	4.29	0.18	18.15	0.58
Cirrhosis	0.50	0.06	0.206	0.034	22.08	1.76	64.20	5.16
PUD	0.50	0.05	0.223	0.025	22.72	1.76	61.84	7.63
UC	0.49	0.06	0.230	0.034	22.30	1.76	62.76	7.49
IBS	0.51	0.06	0.221	0.030	22.62	1.89	63.41	8.47
F value	135.116		71.706		290.441		203.651	
P value	< 0.001		< 0.001		< 0.001		< 0.001	

Table 5. Effect of rutile and antibiotics on pyruvate and hexokinase

	Pyruvate % change (Increase with Rutile)		Pyruvate % change (Decrease with Doxy)		Hexokinase % change (Increase with Rutile)		Hexokinase % change (Decrease with Doxy)	
Normal	4.34	0.21	18.43	0.82	4.21	0.16	18.56	0.76
Cirrhosis	21.52	2.26	60.42	7.65	21.70	1.90	65.26	5.62
PUD	21.29	2.38	57.56	8.70	22.80	2.33	64.43	5.74
UC	21.34	2.24	60.25	8.94	22.29	2.22	65.14	5.66
IBS	20.74	1.47	61.98	6.44	22.36	2.40	63.46	5.69
F value	321.255		115.242		292.065		317.966	
P value	< 0.001		< 0.001		< 0.001		< 0.001	

Table 6. Effect of rutile and antibiotics on hydrogen peroxide and delta amino levulinic acid

	H_2O_2 % (Increase with Rutile)		H_2O_2 % (Decrease with Doxy)		ALA % (Increase with Rutile)		ALA % (Decrease with Doxy)	
Normal	4.43	0.19	18.13	0.63	4.40	0.10	18.48	0.39
Cirrhosis	23.46	1.61	61.77	6.79	23.98	1.72	66.76	4.01
PUD	22.38	1.65	64.59	7.12	23.52	1.74	67.75	3.43
UC	23.65	1.11	59.37	6.93	23.13	1.96	65.86	3.83
IBS	23.22	1.76	59.12	5.14	23.32	1.95	66.69	3.91
F value	380.721		171.228		372.716		556.411	
P value	< 0.001		< 0.001		< 0.001		< 0.001	

Table 7. Effect of rutile and antibiotics on ATP synthase and cytochrome F 420

	ATP synthase % (Increase with Rutile)		ATP synthase % (Decrease with Doxy)		CYT F420 % (Increase with Rutile)		CYT F420 % (Decrease with Doxy)	
Normal	4.40	0.11	18.78	0.11	4.48	0.15	18.24	0.66
Cirrhosis	23.27	1.56	66.43	3.77	22.46	2.39	61.42	7.26
PUD	23.09	1.43	66.43	4.07	22.41	2.02	60.47	8.32
UC	23.14	1.80	66.40	3.64	22.95	1.53	58.86	6.97
IBS	23.16	1.31	67.28	3.54	22.52	1.33	61.43	11.16
F value	449.503		673.081		306.749		130.054	
P value	< 0.001		< 0.001		< 0.001		< 0.001	

The RNA viroids and its complementary DNA developed into cholesterol enveloped RNA and DNA viruses like herpes, retrovirus, influenza virus, borna virus, cytomegalo virus and Ebstein Barr virus by recombining with eukaryotic and human genes resulting in viral speciation. Bacterial and viral species are ill defined and fuzzy with all of them forming one common genetic pool with frequent horizontal gene transfer and recombination. Thus the multi and unicellular eukaryote with its genes serves the purpose of prokaryotic and viral speciation. The multicellular eukaryote developed so that their endosymbiotic archaeal colonies could survive and forage better. The multicellular eukaryotes are like bacterial biofilms. The archaea and bacteria with a mevalonate pathway uses the extracellular RNA viroids and DNA viroids for quorum sensing and in the generation of symbiotic biofilm like structures which develop into multicellular eukaryotes [26,27]. The endosymbiotic archaea and bacteria with mevalonate pathway still uses the RNA viroids and DNA viroids for the regulation of muticellular eukaryote. Pollution is induced by the primitive nanoarchaea and mevalonate pathway bacteria synthesised PAH and methane leading on to redox stress. Redox stress leads to sodium potassium ATPase inhibition, inward movement of plasma membrane cholesterol, defective SREBP sensing, increased cholesterol synthesis and nanoarchaeal/mevalonate pathway bacterial growth [28]. Redox stress leads on to viroidal and archaeal multiplication. Redox stress can also lead to HERV reverse transcriptase and integrase expression. The noncoding DNA is formed of integrating RNA viroidal complementary DNA and archaea with the integration going on as a continuing event. The archaeal pox like dsDNA virus forms evolutionarily the nucleus. The integrated viroidal, archaeal and mevalonate pathway bacterial sequences can undergo vertical transmission and can exist as genomic parasites. The genomic integrated archaea, mevalonate pathway bacteria and viroids form a genomic reserve of bacteria and viruses which can recombine with human and eukaryotic genes producing bacterial and viral speciation. Bacteria and viruses have been related to the pathogenesis of cirrhosis liver, ulcerative colitis, irritable bowel syndrome and peptic ulcer disease [29,30]. Helicobacter pylori has been related to the pathogenesis of peptic ulcer disease [29]. Mollicutes, atypical mycobacteria and enterobacteria has been implicated in inflammatory bowel disease [29,30]. Gut bacteria and endotoxinemia contributes to the pathogenesis of cirrhosis liver [29]. Gut bacteria also plays a role in irritable bowel syndrome [29]. The change in the length and grammar of the noncoding region produces eukaryotic speciation and individuality [31]. Changes in the length of noncoding region especially human endogenous retroviruses can lead onto autoimmune diseases [32]. The integration of nanoarchaea, mevalonate pathway prokaryotes and viroids in to the eukaryotic and human genome produces a chimera which can multiply producing biofilm like multicellular structures having a mixed archaeal, viroidal, prokaryotic and eukaryotic characters which is a regression from the multicellular eukaryotic tissue This results in a new neuronal, metabolic, immune and tissue phenotype or microchimeras leading to human diseases like cirrhosis liver, ulcerative colitis, irritable bowel syndrome and peptic ulcer disease. The microchimeras formed can lead to autoantigens, immune activation and autoimmune pathology. Autoimmunity has been described in inflammatory bowel disease [29].

Archaea and RNA viroid can bind the TLR receptor induce NFKB producing immune activation and cytokine TNF alpha secretion. The archaeal DXP and mevalonate pathway metabolites can bind γδ TCR and digoxin induced calcium signalling can activate NFKB producing chronic immune activation [2,33]. The archaeal cholesterol aromatase generated PAH can produce immune activation. The archaea and viroid induced chronic immune activation and generation of superantigens can lead on to autoimmune disease and immune activation. Immune activation has been related to the pathogenesis of cirrhosis liver, ulcerative colitis, irritable bowel syndrome and peptic ulcer disease [29,30].

The archaea and viroids can regulate the nervous system including the NMDA synaptic transmission[2]. NMDA can be activated by digoxin induced calcium oscillations, PAH and viroid induced RNA interference [2]. The cholesterol ring oxidase generated pyruvate can be converted by the GABA shunt pathway to glutamate. The archaeal cholesterol aromatase can generate serotonin [17]. Glutamatergic and serotoninergic transmission can lead to immune activation which is important in the pathogenesis of cirrhosis liver, ulcerative colitis, irritable bowel syndrome and peptic ulcer disease. Monoamine neurotransmitters and glutamate have been implicated in abnormal gut motility of irritable bowel syndrome. The higher degree of integration of the archaea into the genome produces increased digoxin synthesis producing right hemispheric dominance and lesser degree producing left hemispheric dominance [2]. Right hemispheric dominance can lead to cirrhosis liver, ulcerative colitis, irritable bowel syndrome and peptic ulcer disease [2].

Archaea, viroids and digoxin can induce the host AKT PI3K, AMPK, HIF alpha and NFKB producing the Warburg metabolic phenotype [34]. The increased glycolytic hexokinase activity, decrease in blood ATP, leakage of cytochrome C, increase in serum pyruvate and decrease in acetyl CoA indicates the generation of the Warburg phenotype. There is induction of glycolysis, inhibition of PDH activity and mitochondrial dysfunction resulting in inefficient energetics. The lymphocytes depend on glycolysis for their energy needs. The increased glycolysis induced by the Warburg phenotype leads to immune activation.

Lactic acid generated by increased glycolysis leads to immune stimulation. Immune activation as noted before is important in the pathogenesis of cirrhosis liver, ulcerative colitis, irritable bowel syndrome and peptic ulcer disease. Cholesterol oxidase activity, increased glycolysis related NADPH oxidase activity, bacterial porphyrin induced redox stress and mitochondrial dysfunction generates free radicals important in the pathogenesis of cirrhosis liver, ulcerative colitis, irritable bowel syndrome and peptic ulcer disease. The accumulated pyruvate enters the gaba shunt pathway and is converted to citrate which is acted upon by citrate lyase and converted to acetyl CoA, used for cholesterol synthesis [34].

The pyruvate can be converted to glutamate and ammonia which is oxidised by archaea for energy needs. The increased cholesterol substrate leads to increased archaeal growth and digoxin synthesis leading to metabolic channelling to the mevalonate pathway. The archaeal cholesterol catabolism can deplete the lymphocytic cell membranes of cholesterol resulting in alteration of lymphocytic cell membrane microdomains related receptors producing immune activation. Hyperdigoxinemia is important in the pathogenesis of cirrhosis liver, ulcerative colitis, irritable bowel syndrome and peptic ulcer disease [2]. Digoxin can increase lymphocytic intracellular calcium which leads on to induction of NFKB and immune activation [2].

The archaeal bile acids can bind GPCR and modulate D2 regulating the conversion of T4 to T3. T3 activates uncoupling proteins reducing redox stress. Bile acids can also activate NRF ½ inducing NQO1, GST, HOI reducing redox stress. Bile acids can bind PXR inducing the bile acid shunt pathway of cholesterol detoxification. Bile acids can bind macrophage GPCR and VDR producing immunosuppression and inhibiting NFKB. This helps to modulate the archaea and viroid induced chronic immune activation. Bile acids are thus protective compounds and put a break on the archaea and viroid induced changes [35].

Thus the actinide, viroid and mevalonate pathway bacteria induced metabolic, genetic, immune and neuronal transmission changes can lead onto cirrhosis liver, ulcerative colitis, irritable bowel syndrome and peptic ulcer disease.

References

[1] Valiathan M.S., Somers, K., Kartha, C.C. (1993). *Endomyocardial Fibrosis.* Delhi: Oxford University Press.

[2] Kurup R., Kurup, P.A. (2009). *Hypothalamic digoxin, cerebral dominance and brain function in health and diseases.* New York: Nova Science Publishers.

[3] Hanold D., Randies, J.W. (1991). Coconut cadang-cadang disease and its viroid agent, *Plant Disease,* 75, 330-335.

[4] Edwin B.T., Mohankumaran, C. (2007). Kerala wilt disease phytoplasma: Phylogenetic analysis and identification of a vector, *Proutista moesta, Physiological and Molecular Plant Pathology,* 71(1-3), 41-47.

[5] Eckburg P.B., Lepp, P.W., Relman, D.A. (2003). Archaea and their potential role in human disease, *Infect. Immun,* 71, 591-596.

[6] Adam Z. (2007). Actinides and Life's Origins, *Astrobiology,* 7, 6-10.

[7] Schoner W. (2002). Endogenous cardiac glycosides, a new class of steroid hormones, *Eur. J. Biochem.,* 269, 2440-2448.

[8] Davies P.C.W., Benner, S.A., Cleland, C.E., Lineweaver, C.H., McKay, C.P., Wolfe-Simon, F. (2009). Signatures of a Shadow Biosphere, *Astrobiology,* 10, 241-249.

[9] Richmond W. (1973). Preparation and properties of a cholesterol oxidase from nocardia species and its application to the enzymatic assay of total cholesterol in serum, *Clin. Chem.,* 19, 1350-1356.

[10] Snell E.D., Snell, C.T. (1961). *Colorimetric Methods of Analysis.* Vol 3A. New York: Van NoStrand.

[11] Glick D. (1971). *Methods of Biochemical Analysis.* Vol 5. New York: Interscience Publishers.

[12] Colowick, Kaplan, N.O. (1955). *Methods in Enzymology.* Vol 2. New York: Academic Press.

[13] Maarten A.H., Marie-Jose, M., Cornelia, G., van Helden–Meewsen, Fritz, E., Marten, P.H. (1995). Detection of muramic acid in human spleen, *Infection and Immunity,* 63(5), 1652 – 1657.

[14] Smit A., Mushegian, A. (2000). Biosynthesis of isoprenoids via mevalonate in Archaea: the lost pathway, *Genome Res.,* 10(10), 1468-84.

[15] Van der Geize R., Yam, K., Heuser, T., Wilbrink, M.H., Hara, H., Anderton, M.C. (2007). A gene cluster encoding cholesterol catabolism in a soil actinomycete provides insight into Mycobacterium tuberculosis survival in macrophages, *Proc. Natl. Acad. Sci. USA,* 104(6), 1947-52.

[16] Francis A.J. (1998). Biotransformation of uranium and other actinides in radioactive wastes, *Journal of Alloys and Compounds,* 271(273), 78-84.

[17] Probian C., Wülfing, A., Harder, J. (2003). Anaerobic mineralization of quaternary carbon atoms: Isolation of denitrifying bacteria on pivalic acid (2,2-Dimethylpropionic acid), *Applied and Environmental Microbiology,* 69(3), 1866-1870.

[18] Vainshtein M., Suzina, N., Kudryashova, E., Ariskina, E. (2002). New Magnet-Sensitive Structures in Bacterial and Archaeal Cells, *Biol. Cell,* 94(1), 29-35.

[19] Tsagris E.M., de Alba, A.E., Gozmanova, M., Kalantidis, K. (2008). Viroids, *Cell Microbiol.,* 10, 2168.

[20] Horie M., Honda, T., Suzuki, Y., Kobayashi, Y., Daito, T., Oshida, T. (2010). Endogenous non-retroviral RNA virus elements in mammalian genomes, *Nature,* 463, 84-87.

[21] Hecht M., Nitz, N., Araujo, P., Sousa, A., Rosa, A., Gomes, D. (2010). Genes from Chagas parasite can transfer to humans and be passed on to children. Inheritance of DNA Transferred from American Trypanosomes to Human Hosts, *PLoS ONE,* 5, 2-10.

[22] Flam F. (1994). Hints of a language in junk DNA, *Science,* 266, 1320.

[23] Horbach S., Sahm, H., Welle, R. (1993). Isoprenoid biosynthesis in bacteria: two different pathways? *FEMS Microbiol. Lett,* 111, 135–140.

[24] Gupta R.S. (1998). Protein phylogenetics and signature sequences: a reappraisal of evolutionary relationship among archaebacteria, eubacteria, and eukaryotes, *Microbiol. Mol. Biol. Rev.,* 62, 1435–1491.

[25] Hanage W., Fraser, C., Spratt, B. (2005). Fuzzy species among recombinogenic bacteria, *BMC Biology,* 3, 6-10.

[26] Webb J.S., Givskov, M., Kjelleberg, S. (2003). Bacterial biofilms: prokaryotic adventures in multicellularity, *Curr. Opin. Microbiol.,* 6(6), 578–85.

[27] Whitchurch C.B., Tolker-Nielsen, T., Ragas, P.C., Mattick, J.S. (2002). Extracellular DNA Required for Bacterial Biofilm Formation. *Science,* 295(5559), 1487.

[28] Chen Y., Cai, T., Wang, H., Li, Z., Loreaux, E., Lingrel, J.B. (2009). Regulation of intracellular cholesterol distribution by Na/K-ATPase, *J. Biol. Chem.,* 284(22), 14881-90.

[29] Fauci A.S. (2009). Harrison's Principles of Internal Medicine. New York: McGraw-Hill.

[30] Wirostko E., Johnson, L., Wirostko, B. (1990). Ulcerative colitis associated chronic uveitis. Parasitization of intraocular leucocytes by mollicute-like organisms, *J. Submicrosc. Cytol. Pathol.,* 22(2), 231-9.

[31] Poole A.M. (2006). Did group II intron proliferation in an endosymbiont-bearing archaeon create eukaryotes? *Biol. Direct,* 1, 36-40.

[32] Villarreal L.P. (2006). How viruses shape the tree of life, *Future Virology,* 1(5), 587-595.

[33] Eberl M., Hintz, M., Reichenberg, A., Kollas, A., Wiesner, J., Jomaa, H. (2010). Microbial isoprenoid biosynthesis and human $\gamma\delta$ T cell activation, *FEBS Letters,* 544(1), 4-10.

[34] Wallace D.C. (2005). Mitochondria and Cancer: Warburg Addressed, *Cold Spring Harbor Symposia on Quantitative Biology,* 70, 363-374.

[35] Lefebvre P., Cariou, B., Lien, F., Kuipers, F., Staels, B. (2009). Role of Bile Acids and Bile Acid Receptors in Metabolic Regulation, *Physiol Rev,* 89(1), 147 - 191.

Index

A

B

C

D

E

F

G

H

I

J

K

L

M

N

O

P

Q

R

S

T

U

V

W